Forest Management, Conflict and Social-Ecological Systems in a Changing World

Forest Management, Conflict and Social-Ecological Systems in a Changing World

Editors

Juan F. Fernández-Manjarrés
Roxane Sansilvestri

MDPI • Basel • Beijing • Wuhan • Barcelona • Belgrade • Manchester • Tokyo • Cluj • Tianjin

Editors
Juan F. Fernández-Manjarrés
Université Paris Saclay
France

Roxane Sansilvestri
University of Alcala de Henares
Spain

Editorial Office
MDPI
St. Alban-Anlage 66
4052 Basel, Switzerland

This is a reprint of articles from the Special Issue published online in the open access journal *Forests* (ISSN 1999-4907) (available at: https://www.mdpi.com/journal/forests/special_issues/Forest_Management).

For citation purposes, cite each article independently as indicated on the article page online and as indicated below:

LastName, A.A.; LastName, B.B.; LastName, C.C. Article Title. *Journal Name* **Year**, *Volume Number*, Page Range.

ISBN 978-3-0365-3379-7 (Hbk)
ISBN 978-3-0365-3380-3 (PDF)

Cover image courtesy of Juan Fernández-Manjarrés

Contents

About the Editors

Juan Fernández-Manjarrés (Senior Scientist at the Natinal Research Center CNRS in France) is an ecologist addressing interdisciplinary issues between ecology and the social sciences. His current interests include studying the role of agroforestry systems in Equatorial America and Western Europe, especially with respect to adaptation and early warnings of multiple risks.

Roxane Sansilvestri (Postdoctoral researcher at the University of Alcalá de Henares) is a socioecologist, researching how actors and institutions organize themselves to face complex adaptation problems, specifically in forest socioecological systems. Her particular interests lie in value and perception issues, and adaptive capacity in crisis contexts.

Preface to "Forest Management, Conflict and Social-Ecological Systems in a Changing World"

Biomedical sensing and imaging technology are of great significance for the early detection, rapid diagnosis and precise treatment of diseases. In the last three decades, considerable research efforts have been devoted to biomedical sensors and imaging systems. Biomedical sensing and imaging systems can be roughly divided into optical imaging technology, electrical imaging technology, acoustic imaging technology, radiographic imaging technology, etc. As each technology has its own unique advantages, various new types of sensing and monitoring equipment have been invented, and improving the performance of the existing equipment is essential. In recent years, biomedical sensing and imaging have been applied in many aspects of biomedical research, such as biomolecular testing, health monitoring, and disease diagnosis and treatment. Biosensors convert biomedical signals to electrical signals for further acquisition and processing using downstream devices and algorithms, which are also crucial for rapid and accurate biomedical diagnosis. At present, biosensor design emphasizes portability, networked detection, integration and intelligentization. Moreover, research methods are being developed regarding multi-modality and multi-function fusion, which has become a trend in the development of biological imaging systems. Some multi-modal imaging equipment has been successfully used in clinical applications, which can provide more accurate and high-resolution anatomical, functional and biochemical information. With the development of computer technology, physics, molecular biology, materials chemistry and other disciplines, new software and hardware devices have been developed, opening up a new world for the development and application of biological imaging technology. This book, which is a Special Issue of Biomedical Sensing and Imaging, aims to collect recent advances in the above topics regarding biosensing and imaging, especially in wearable/smart biosensors and advanced imaging algorithms.

Juan F. Fernández-Manjarrés, Roxane Sansilvestri
Editors

forests MDPI

Editorial

Forest Management, Conflict and Social–Ecological Systems in a Changing World

Juan F. Fernández-Manjarrés [1,*], Josephine MacHunter [2] and Miguel A. Zavala [3,4]

1 Ecologie Systématique Evolution, CNRS, AgroParisTech, Université Paris-Saclay, 91405 Orsay, France
2 Arthur Rylah Institute for Environmental Research, Department of Environment, Heidelberg, VIC 3084, Australia; josephine.machunter@delwp.vic.gov.au
3 Grupo de Ecología y Restauración Forestal (FORECO), Departamento de Ciencias de la Vida, Universidad de Alcalá, Alcalá de Henares, 28805 Madrid, Spain; ma.zavala@uah.es
4 Instituto Franklin, Universidad de Alcalá, Calle Trinidad 1, 28801 Alcalá de Henares, Spain
* Correspondence: juan.fernandez@universite-paris-saclay.fr

Citation: Fernández-Manjarrés, J.F.; MacHunter, J.; Zavala, M.A. Forest Management, Conflict and Social–Ecological Systems in a Changing World. *Forests* **2021**, *12*, 1459. https://doi.org/10.3390/f12111459

Received: 28 September 2021
Accepted: 22 October 2021
Published: 26 October 2021

1. Introduction

Conflicts are ubiquitous in forest management because of several overlapping temporal and spatial issues (see examples in tropical and temperate areas [1–5]). First, the long temporal cycles of forests make decision making particularly difficult, as decisions made today can have impacts for centuries to come [6]. Second, the large spatial scales inherent to forest ecosystems imply both a large number of ecosystem-specific management plans, but also a wide variety of actors. Finally, because of the long cycles, management plans (or the lack of them) are inherited often from previous generations, and new actors may not agree with the decisions taken before.

Forests are, par excellence, multifunctional ecosystems. By multifunctional, we mean that forests either currently provide or have the potential to provide multiple benefits or ecosystem services for society [7,8]. However, this multifunctionality comes at a cost because forests that occur as social–ecological systems (SESs, also known as coupled human or natural systems, or CHANS) can rarely be optimized to simultaneously achieve all desirable outcomes, necessitating trade-offs whose acceptance differs among interest groups [9]. Multifunctionality is a concept that has made it to management programs and public policies in most parts of the world. For instance, in many countries of Europe, it has been explicitly included in the legislation regarding public and private forests since the mid-1950s [10,11] and it is a current management framework in countries such as India [12] and Brazil [13], to cite just a couple. Behind this concept, what is really at stake are notions of shared landscape. As such, conflicts related to multifunctionality are frequently related to land-sparing or land-sharing approaches, a debate that is far from being closed [14]. The multifunctionality paradigm competes today with new ideas, such as rewilding [15], which, in turn, can accentuate lingering conflicts if not properly handled [16].

If conflicts are ubiquitous, as we state above, inferring general tendencies may be an important advancement towards a forest SES theory development. The question that follows then is if forest SESs would have particular types of conflicts not found in other ecosystems. Another particularity of forests is that, even if they occur on private lands, they may be under the control or strict supervision of governments in most countries of the world. The same applies to forests occurring in first nation or indigenous reserves, in which, despite the autonomy granted, the management should be in accord with national regulations. Here, our main premise would be that forest SESs would be prone to conflict due to lagged responses between the time that decisions are made by local actors and changes in the ecological and social systems, which do not necessarily follow the same cycles of decision making of local institutions, regional forest authorities, and national governments. These lags in decision making are further confounded by different views of the individual actors and regional administrations that understand governance differently. In this editorial article, we present first a general typology of sources of conflicts before introducing the contributions by the authors on different topics, and we conclude by raising awareness of the effects that current "transition" approaches may have for forest management considered as SESs.

2. A Simplified Typology

To assist with exploration of the origin of the tensions, we propose a simple two-dimension typology that integrates a number of the conflicts observed in forest SESs (Figure 1).

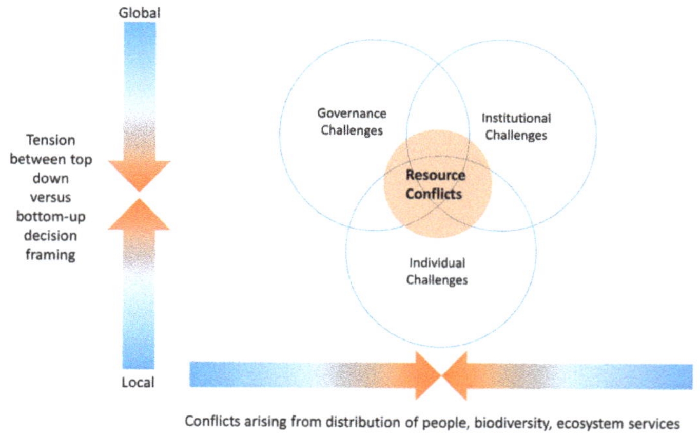

Figure 1. A simplified typology of some of the possible sources of conflicts in forests. The vertical axis represents the newly created pressures of international agreements that impact directly standing forests, afforestation, and reforestation programs. The horizontal axis represents the conflicts caused by the distribution of people (including property rights), the structure and composition of forests, and the different ecosystem services people are expecting to find. At the cross-roads of these vertical and horizontal dimensions, challenges arising from institutional factors, disputes in governance, and individual cognitive biases can collide, creating different conflicts at different scales. Transition policies (i.e., increased use of wood biomass for heating) often occur in a top-down fashion from national commitments and international agreements. Local knowledge would be related to the individual challenges regarding what management options are acceptable or not.

In many instances, conflicts arise not because of unwillingness of the parts, but because of inherent problems in the process of decision making (circle intersections in Figure 1). Institutional challenges include issues with data where researchers, managers, and decision makers facing multiple sources of uncertainty (for a review, see [17]), which are often overlooked in the decision making process [18]. Often, the decision frame is poorly articulated or may neglect key concerns of the actors involved, resulting in governance issues arising from mismatched expectations about what a desirable outcome looks like over relevant spatial and temporal scales. The overarching values to be achieved through candidate management interventions may be poorly articulated or lack discrete, measurable objectives to facilitate the evaluation of options. Often, the management options themselves may reflect status quo traps (or other individual biases) rather than genuine exploration of novel approaches that may better support attainment of desirable outcomes [19,20]. In the next section, we discuss how contributions fit in the typology, and at what level decision making is contributing to solving, or else perpetuating, the different described situations.

3. Some Issues Raised by the Contributions to the Special Issue

3.1. Conflicts Arising from Climate Adaptation and Mitigation Programs

Tackling the effects of climate change on SES and preventing further climate change is a policy and management target that will arguably remain for decades to come. In the case of forests, early approaches focused on the role of forest ecosystems as stabilizers of local and regional climates, and, of course, as carbon sinks. For instance, recent findings point out that fast-growing secondary forest in the Amazon basin is more efficient as a carbon sink compared to old growth forests in the region [21]. On the other hand, quantifying whether old growth forests are carbon neutral or sinks is a subject of an ongoing debate [22,23] of far-reaching consequences. As countries continue to engage in carbon emission/sequestration targets, local management will be impacted by policies that have a top-down origin, not only from a national level, but from an international level too, often labeled as "transitions" [24].

In a way, this category of conflicts exhibits most, if not all, of the dimensions depicted in Figure 1, as top-down policies collide with local expectations of actors, which themselves have other unresolved issues regarding access to markets, unfair competition, or simply rights to use the forest resource itself.

In this Special Issue, Royer-Tardiff and colleagues [25] present a portfolio approach for implementing forest zonings, in which zones that are more conserved and those that are more impacted can have different climate change adaptation roles. For the latter, the authors propose that, in heavily impacted areas where restoration is unlikely, new afforestation strategies could be tested without entering into competition with areas less exposed to climate change and that can serve as biodiversity reservoirs. The example in Canadian boreal forests that they provide of functional zoning aims at distributing in the landscape conflicting land uses, an approach that would be akin to a mixed land-sparing/land-sharing strategy. Akita and Ohe [26], in this issue, propose the use of carbon credits to avoid the lack of management in former production forests in Japan that have become nonprofitable if commercialized as timber. In this case, carbon credits act as an innovation tool and can act as way of avoiding land use conflicts, as funds for management can be sourced through this alternative. Another contribution to this journal issue (Fouqueray and colleagues) also considers the roles of carbon credits within concertation by different actors around biodiversity and production issues.

Fouqueray and colleagues [27] analyze an ongoing strategy in French forests that provides subsidies for management programs seeking to increase carbon sequestration. As this strategy can be interpreted as promoting homogeneous, low diversity, low quality managed forests, the analyzed programs have a social–ecological approach, in which consensus is sought with local partners to avoid conflicts before the programs are deployed. Additionally, in France, Sansilvestri and colleagues [28] show in high detail how top-down European policies of energetic transition model the use of wood biomass following structures inherited from fossil fuel economies that are not in accord with the way actors are organized around the forest resources.

On the other hand, conflicts in tropical countries regarding adaptation and mitigation issues may be more difficult to solve. Alusiola and colleagues, in this issue [29], analyze through the lens of political ecology how conflicts have arisen in reducing emissions from deforestation and forest degradation in developing countries (REDD+) programs in selected case studies located in East Africa (Ethiopia, Uganda, Tanzania), Southeast Asia (Indonesia, Vietnam), and Central America (Panama). The good intentions of these initiatives are smeared by unexpected side effects that include increased injustices and restrictions over forest resources, and the aggravation of historic land tenure conflicts, among others. Restrictions over forest access are one of the issues that we depict in the horizontal axis of the typology in Figure 1, as they are influenced not only by the spatial distribution of property in the forests, but by interactions of actors that may try to impose conflicting governance approaches.

3.2. Protected Areas and Biodiversity

Conflicting goals between biodiversity and forestry activities are one of the more common issues worldwide. In this issue, Shneider and colleagues [30] use archival research coupled with stakeholder interviews to gain insights into a recurring problem of land preservation and forestry zones in the Czech Republic. In this case, they analyzed the perception of the local society that considers that the declaration of protected areas is a very restrictive instrument. In this study, the process started as a top-down measure from the central government that could not reach consensus with local stakeholders, leading to an abandonment of the project. Interestingly, the results refuted or did not confirm most of the arguments of stakeholders against the declaration of the protected area, showing how a lack of communication can feed unwarranted bias of the public, a situation we depict on the intersection of circles representing personal, governance, and institutional issues in Figure 1. This study is also an example of the horizontal issues that we address in the typology, as the spatial distribution of biodiversity is a "horizontal" dimension highly impacted by property rights and/or governance over forests that require zoning to define protection and extraction areas.

3.3. Loss of Local Knowledge

In the literature, calls for more holistic management approaches are increasing [31]. Holistic management seeks to avoid conceptual and practical oppositions in terms of private/public lands, conservation/exploitation, and academic knowledge/local knowledge. Among these, local knowledge and local indigenous knowledge consideration appear to be a way of avoiding conflicts and resolving opposing positions. In this issue, Branca and colleagues [32] reconstruct through interviews the historical–cultural and social cohesion function that the forest plays in a rural community in the Mediterranean island of Sardinia. Rural population abandonment, with the concomitant loss of local knowledge, was identified as the main risk facing the management of coppiced forests. New propositions of shorter rotation times sparked a debate that hit the national press in Italy, labeling the plan as the destruction of a millennium-old forest mostly by an urban public not familiar with traditional coppice methods. The example by Branca and colleagues adds to an ever-growing body of research demonstrating that concertation and incorporation of local knowledge is a condition for sustainable management of forests, for both tropical and temperate countries [33]. Local knowledge would be represented in our typology within the circle depicting individual challenges, because local actors have to make decisions based on what they consider right for their ecosystem and what is legislated by local authorities.

3.4. Telecoupling, Local Resilience, and Use Conflicts

In the Mediterranean basin, humans have used forests for centuries, with variable outcomes, ranging from degradation to sustainable human cultural landscapes. In this issue, Moreno and colleagues [34] analyze the social–ecological trajectory in a marginal dry-edge maritime pine forest in central Spain, an ecosystem highly vulnerable to desertification. These authors show that, despite recurring impacts of teleconnections (external climatic drivers) and telecouplings (wars, markets, major political changes, etc.), a century old management plan has provided resilience to the ecological and social components of the system, because forest function, including regeneration, and social needs were identified and well integrated by forest authorities. Hence, neglecting forest management in favor of more strict conservation policies could be, in some regions, counterproductive for community and ecosystem resilience.

Various contributing papers account for the impact of telecoupling that is represented in top-down policies, such as the European Union targets for greener energy [28], REDD+ programs [29], and carbon credits [26,27], highlighting the vertical dimension of conflict origin, as depicted in Figure 1.

4. Concluding Remarks: A Cautionary Tale on "Transitions"

Currently, public policies are abound with all sorts of social transition objectives, and research is following with projects, publications, and journals devoted specifically to one or more transitions. As we briefly described above, forests are at the center of climate change adaptation and mitigation, with increased attention given today to new energy sources, such as wind farms or massive afforestation, for energy that will help an energetic transition [22]. However, these new land uses will necessarily imply new trade-offs and challenges. Conflicts will continue to arise if the public and stakeholders observe multifunctional forests replaced by crop trees [35] in the name of an energy transition policy, making this strategy a clear source of considerable problems [36]. There are limits to the sustainability and resilience of forest ecosystems, and we need to carefully consider how to cope with these trade-offs, and which actions we undertake with our extant forests, as lagged responses may cause our mistakes to linger for generations to come [6].

Author Contributions: Conceptualization, J.F.F.-M., J.M. and M.A.Z.; writing—original draft preparation, J.F.F.-M.; writing—review and editing, J.F.F.-M., J.M. and M.A.Z. All authors have read and agreed to the published version of the manuscript.

Funding: J.F.F.-M. received support from a grant overseen by the French National Research Agency (ANR) as part of the "Investments d'Avenir" Programme (LabEx BASC; ANR-11-LABX-0034); M.A.Z was supported by grant DARE; RTI2018-096884-B-C32 (Ministerio de Ciencia e Innovación-MICINN, Spain).

Data Availability Statement: Not applicable.

Acknowledgments: The authors would like to thank the Editorial Office of *Forests* for their continuous support and invaluable help for achieving this Special Issue. We would also like to thank Lindy F. Lumsden for a thorough review of the final version of the manuscript.

Conflicts of Interest: The authors declare no conflict of interest.

References

1. Morford, S.; Parker, D.; Rogers, H.; Salituro, C.; Waldichuk, T. Culture, worldviews, communication styles, and conflict in forest management. *J. Ecosyst. Manag.* **2003**, *3*, 3–7.
2. Poynton, S.; Hurditch, W. Cooperation and conflict in forest management: Applying a theoretical model to the Australian problem. *Aust. For.* **1995**, *58*, 58–64. [CrossRef]
3. Pravat, P.S.; Humphreys, D. Using a multilevel approach to analyse the case of forest conflicts in the Terai, Nepal. *For. Policy Econ.* **2013**, *33*, 47–55. [CrossRef]
4. De Jong, W.; Ruiz, S.; Becker, M. Conflicts and communal forest management in northern Bolivia. *For. Policy Econ.* **2006**, *8*, 447–457. [CrossRef]
5. Blicharska, M.; Angelstam, P.; Giessen, L.; Hilszczański, J.; Hermanowicz, E.; Holeksa, J.; Jacobsen, J.; Jaroszewicz, B.; Konczal, A.; Konieczny, A. Between biodiversity conservation and sustainable forest management—A multidisciplinary assessment of the emblematic Białowieża Forest case. *Biol. Conserv.* **2020**, *248*, 108614. [CrossRef]
6. Fernandez-Manjarres, J.F.; Tschanz, L. Assisted Colonization: Protect Managed Forests. *Science* **2010**, *330*, 1318. [CrossRef]
7. Byrnes, J.E.; Gamfeldt, L.; Isbell, F.; Lefcheck, J.S.; Griffin, J.N.; Hector, A.; Cardinale, B.J.; Hooper, D.U.; Dee, L.E.; Emmett Duffy, J. Investigating the relationship between biodiversity and ecosystem multifunctionality: Challenges and solutions. *Methods Ecol. Evol.* **2014**, *5*, 111–124. [CrossRef]
8. Mastrangelo, M.E.; Weyland, F.; Villarino, S.H.; Barral, M.P.; Nahuelhual, L.; Laterra, P. Concepts and methods for landscape multifunctionality and a unifying framework based on ecosystem services. *Landsc. Ecol.* **2014**, *29*, 345–358. [CrossRef]
9. Lu, N.; Liu, L.; Yu, D.; Fu, B. Navigating trade-offs in the social-ecological systems. *Curr. Opin. Environ. Sustain.* **2021**, *48*, 77–84. [CrossRef]
10. Borrass, L.; Kleinschmit, D.; Winkel, G. The "German model" of integrative multifunctional forest management—Analysing the emergence and political evolution of a forest management concept. *For. Policy Econ.* **2017**, *77*, 16–23. [CrossRef]
11. Bončina, A.; Simončič, T.; Rosset, C. Assessment of the concept of forest functions in Central European forestry. *Environ. Sci. Policy* **2019**, *99*, 123–135. [CrossRef]
12. Rai, A.; Singh, A.K.; Mehrotra, S.; Singh, N. Multifunctional tropical dry forests systems of India: Current need and future directions. In *Innovation of Life Science Research*; NOVA Publisher: Hauppauge, NY, USA, 2019; pp. 373–399.
13. Sardeshpande, M.; Shackleton, C. Wild edible fruits: A systematic review of an under-researched multifunctional NTFP (non-timber forest product). *Forests* **2019**, *10*, 467. [CrossRef]

14. Grass, I.; Loos, J.; Baensch, S.; Batáry, P.; Librán-Embid, F.; Ficiciyan, A.; Klaus, F.; Riechers, M.; Rosa, J.; Tiede, J.; et al. Land-sharing/-sparing connectivity landscapes for ecosystem services and biodiversity conservation. *People Nat.* **2019**, *1*, 262–272. [CrossRef]
15. Perino, A.; Pereira, H.M.; Navarro, L.M.; Fernández, N.; Bullock, J.M.; Ceaușu, S.; Cortés-Avizanda, A.; van Klink, R.; Kuemmerle, T.; Lomba, A. Rewilding complex ecosystems. *Science* **2019**, *364*, 1328–1348. [CrossRef]
16. Van Meerbeek, K.; Muys, B.; Schowanek, S.D.; Svenning, J.-C. Reconciling conflicting paradigms of biodiversity conservation: Human intervention and rewilding. *BioScience* **2019**, *69*, 997–1007. [CrossRef]
17. Regan, H.M.; Colyvan, M.; Burgman, M.A. A taxonomy and treatment of uncertainty for ecology and conservation biology. *Ecol. Appl.* **2002**, *12*, 618–628. [CrossRef]
18. Hammond, J.S.; Keeney, R.L.; Raiffa, H. *Smart Choices: A Practical Guide to Making Better Decisions*; Harvard Business Review Press: Boston, Mass, USA, 2015; p. 243.
19. Hammond, J.S.; Keeney, R.L.; Raiffa, H. The hidden traps in decision making. *Harv. Bus. Rev.* **1998**, *76*, 47–58. [PubMed]
20. Kahneman, D.; Knetsch, J.L.; Thaler, R.H. Anomalies: The endowment effect, loss aversion, and status quo bias. *J. Econ. Perspect.* **1991**, *5*, 193–206. [CrossRef]
21. Heinrich, V.H.A.; Dalagnol, R.; Cassol, H.L.G.; Rosan, T.M.; de Almeida, C.T.; Silva Junior, C.H.L.; Campanharo, W.A.; House, J.I.; Sitch, S.; Hales, T.C.; et al. Large carbon sink potential of secondary forests in the Brazilian Amazon to mitigate climate change. *Nat. Commun.* **2021**, *12*, 1785. [CrossRef]
22. Luyssaert, S.; Schulze, E.-D.; Börner, A.; Knohl, A.; Hessenmöller, D.; Law, B.E.; Ciais, P.; Grace, J. Old-growth forests as global carbon sinks. *Nature* **2008**, *455*, 213–215. [CrossRef]
23. Gundersen, P.; Thybring, E.E.; Nord-Larsen, T.; Vesterdal, L.; Nadelhoffer, K.J.; Johannsen, V.K. Old-growth forest carbon sinks overestimated. *Nature* **2021**, *591*, E21–E23. [CrossRef]
24. Nik, V.M.; Perera, A. The Importance of Developing Climate-Resilient Pathways for Energy Transition and Climate Change Adaptation. *One Earth* **2020**, *3*, 423–424. [CrossRef]
25. Royer-Tardif, S.; Bauhus, J.; Doyon, F.; Nolet, P.; Thiffault, N.; Aubin, I. Revisiting the Functional Zoning Concept under Climate Change to Expand the Portfolio of Adaptation Options. *Forests* **2021**, *12*, 273. [CrossRef]
26. Akita, N.; Ohe, Y. Sustainable Forest Management Evaluation Using Carbon Credits: From Production to Environmental Forests. *Forests* **2021**, *12*, 1016. [CrossRef]
27. Fouqueray, T.; Génin, L.; Trommetter, M.; Frascaria-Lacoste, N. Efficient, Sustainable, and Multifunctional Carbon Offsetting to Boost Forest Management: A Comparative Case Study. *Forests* **2021**, *12*, 386. [CrossRef]
28. Sansilvestri, R.; Cordier, M.; Lescuyer, T. Winners and Losers in Energy Transition: Study Case of Wood Biomass Power-Plants Implementation in France. *Forests* **2021**, *12*, 1139. [CrossRef]
29. Alusiola, R.A.; Schilling, J.; Klär, P. REDD+ Conflict: Understanding the Pathways between Forest Projects and Social Conflict. *Forests* **2021**, *12*, 748. [CrossRef]
30. Schneider, J.; Ruda, A.; Blahová, M. Stakeholders' Perception of the Impact of the Declaration of New Protected Areas on the Development of the Regions Concerned, Case Study: Czech Republic. *Forests* **2021**, *12*, 580. [CrossRef]
31. Jackson, W.; Freeman, M.; Freeman, B.; Parry-Husbands, H. Reshaping forest management in Australia to provide nature-based solutions to global challenges. *Aust. For.* **2021**, *84*, 50–58. [CrossRef]
32. Branca, G.; Piredda, I.; Scotti, R.; Chessa, L.; Murgia, I.; Ganga, A.; Campus, S.F.; Lovreglio, R.; Guastini, E.; Schwarz, M.; et al. Forest Protection Unifies, Silviculture Divides: A Sociological Analysis of Local Stakeholders' Voices after Coppicing in the Marganai Forest (Sardinia, Italy). *Forests* **2020**, *11*, 708. [CrossRef]
33. Nugroho, K.; Carden, F.; Antlov, H. *Local Knowledge Matters*; Policy Press: Bristol, UK, 2018; p. 174.
34. Moreno-Fernández, D.; Zavala, M.A.; Madrigal-González, J.; Seijo, F. Resilience as a Moving Target: An Evaluation of Last Century Management Strategies in a Dry-Edge Maritime Pine Ecosystem. *Forests* **2021**, *12*, 1151. [CrossRef]
35. Manning, P.; Taylor, G.; Hanley, M. Bioenergy, food production and biodiversity-an unlikely alliance? *Glob. Chang. Biol. Bioenergy* **2014**, 1–18. [CrossRef]
36. Sediri, S.; Trommetter, M.; Frascaria-Lacoste, N.; Fernandez-Manjarrés, J. Transformability as a Wicked Problem: A Cautionary Tale? *Sustainability* **2020**, *12*, 5895. [CrossRef]

forests

MDPI

Article

Resilience as a Moving Target: An Evaluation of Last Century Management Strategies in a Dry-Edge Maritime Pine Ecosystem

Daniel Moreno-Fernández [1,*], Miguel A. Zavala [1], Jaime Madrigal-González [2] and Francisco Seijo [3]

[1] Grupo de Ecología y Restauración Forestal (FORECO), Departamento de Ciencias de la Vida, Universidad de Alcalá, Alcalá de Henares, 28805 Madrid, Spain; ma.zavala@uah.es
[2] Climate Change Impacts and Risks in the Anthropocene, Institute for Environmental Sciences, University of Geneva, 66 Boulevard Carl Vogt, 1205 Geneva, Switzerland; Jaime.MadrigalGonzalez@unige.ch
[3] Instituto de Empresa School of Global and Public Affairs, Calle de María de Molina, 13, 28006 Madrid, Spain; fseijo@faculty.ie.edu
* Correspondence: danielmorenofdez@gmail.com

Abstract: Forests are intrinsically coupled to human dynamics, both temporally and spatially. This evolution is conditioned by global changes in climatic conditions (teleconnections) and distant socio-economical processes (telecoupling). The main goal of this study is to describe the teleconnections and telecoupling dynamics that have shaped structure and processes in a dry-edge—highly vulnerable to desertification—Mediterranean pine forest during the last century and to evaluate the contribution of historical management strategies to this coupled human and natural system's (CHANS) overall resilience. For this study, we collected relevant human and natural system data from a dry edge *Pinus pinaster* Ait. located forest in Central Spain using a CHANS analytical framework operationalizing telecoupling and teleconnection. A key extractive economic activity in the studied forest was resin tapping, which was the main form of land use from the 1920s to the 1950s. Since the 1950s changes in the Spanish economy linked to the emergence of new resin-producing countries, such as China, led to a sharp decline in resin production. Despite additional human system transformations affecting forest governance (e.g., the Spanish Civil War, the transition to democracy, European integration, etc.) and changes in biophysical conditions linked to climate change (e.g., aridification, CO_2 fertilization), the standing stocks of *P. pinaster* increased during the monitoring period due to sound technical and management planning bolstering overall resilience. These historical management decisions, we argue, successfully reconciled overall resilience goals (defined as the maintenance of forest function beyond and desertification avoidance) with three successive historical forest use challenges: intensive firewood collection by local communities in fragile sandy soils, extensive pastoralism in the forest understory and tradeoffs between resin tapping damaged trees, timber production and tree cover as well as the emerging risks of wildfire and climate change.

Keywords: CHANS; globalization; historical data; socio-ecological frameworks; dry-edge

Citation: Moreno-Fernández, D.; Zavala, M.A.; Madrigal-González, J.; Seijo, F. Resilience as a Moving Target: An Evaluation of Last Century Management Strategies in a Dry-Edge Maritime Pine Ecosystem. *Forests* **2021**, *12*, 1151. https://doi.org/10.3390/f12091151

Academic Editor: Laszlo Erdos

Received: 29 June 2021
Accepted: 23 August 2021
Published: 25 August 2021

1. Introduction

Human beings have altered ecosystem structure, function and disturbance throughout history to obtain goods and services from Nature [1,2]. This has been the case for Mediterranean forest ecosystems, which throughout history have provided different functions and services such as wood production [3], carbon storage [4], biodiversity [5], aesthetic functions [6], forage for livestock [7], hunting [8], as well as non-wood products such as resin [9], cork [10] or pine cones [11]. For centuries, humans have exploited Mediterranean forests attempting to maximize economic (e.g., provision of wood, fuelwood, hunting), cultural (e.g., hunting and ecotourism), social (e.g., resources for local communities) and strategic needs (e.g., intervention by the public administration for energy use, water reservoirs, economic production or conservation goals). However, over the last decades, the compounded effects of climate change, human system transformations (e.g., local and international market fluctuations in the demand for forest products) and changes in forest policy governance priorities have pressured the capacity of Mediterranean forests to maintain provision of ecosystem services [12]. Furthermore, as a consequence of changing national and international legislation, new conflicts of interests between stakeholders have emerged which may require changes in local management strategies [13].

These complex social-ecological interactions have been traditionally analyzed through socio-ecological sustainability frameworks [14]. The coupled human and natural systems (CHANS) analytical framework, however, which we apply in this study goes beyond the socio-ecological sustainability approach by focusing on human and natural system feedback couplings, understood as flows, and emergent system properties and how these interactions vary across alternative spatial, temporal and scales [15–17] (we refer readers to the Material and Methods section for further details of the CHANS approach).

The CHANS approach is of particular importance when analyzing ecosystem dynamics as ecosystems change constantly and their functions do not remain in an equilibrium state but rather evolve permanently in time and space. Indeed, ecosystems can cross ecological thresholds or tipping points after which the ecosystem may cease to provide certain functions [18,19]. These spatio-temporal fluctuations may be driven by anthropogenic activities or natural forcings that may interact in complex ways [12,20]. Human activity, through alternative forest management regimes, may modify stand structure and species composition to increase the provision of some functions with lasting legacy effects [12,21,22]. Furthermore, national and international agreements claim to promote afforestation and reforestation programs to enhance the carbon sink capacity of ecosystems, achieve sustainable development goals as well as control soil erosion and regulate the hydric cycle [23,24]. This, in fact, has resulted in land use and CHANS feedback modifications in the targeted territories of developing countries [25,26]. Environmental drivers, such as climate fluctuations or wildfires can interact with biotic drivers, competition, herbivory or pathogens to drive forest dynamics [27,28] and constrain the provision of wood and non-wooden products [29,30].

One of the greatest challenges facing forest managers is climate change, which cannot be analyzed locally since changes in biophysical conditions in one site may influence very distant sites through atmospheric circulation. This phenomenon is known as teleconnection [31]. An additional challenge for current forest managers is economic globalization, which increasingly determines that human land use is also affected by international markets and manufacturing costs [32–34]. In turn, these changes condition local management decisions and may result in tipping points for example in community species (substitution of one forest species with a more lucrative one) [35]. Indeed, the production of an ecosystem good in a given area of the world can be reduced because the manufacturing costs of the by-products are much lower in other areas located thousands of kilometers away [36].

Both human and climatic drivers have both direct and indirect effects on forest dynamics and forest resilience. However, the concept of forest resilience is difficult to define and can be explained through three different, though complementary conceptualizations: engineering, ecological and socio-ecological resilience [37,38]. Whereas engineering and ecological resilience refer to the ability of a system to return to its pre-disturbance stage and the capacity to absorb change and disturbance while maintaining similar feedback dynamics with social or political system variables, respectively [39,40]. Social-ecological resilience, on the other hand, considers the preservation of natural and human system couplings and the adaptive capacity of the entire CHANS [38]. The differences among the three conceptualizations can lead to alternative indicators being employed to quantify or describe forest resilience. According to Nikinmaa et al. [38], the most common indicators in engineering and ecological resilience studies are related to forest structure and biodiversity, while the indicators linked to economic activities as well as financial and technical infrastructure are mainly used to evaluate social-ecological resilience. These authors also argue that the target organization level also varies amongst the different conceptualizations but are nested: that is, engineering resilience (tree level) is nested within ecological resilience (forest level) which is nested within social-ecological resilience (CHANS level).

Due to the complexity of the factors driving the provision of forest ecosystem services, efforts have been made to identify spatial [41,42] and, to a lesser extent, temporal trends [43,44] of multifunctionality. However, data on the effects of distant socio-economic and environmental interactions on the supply of ecosystem services is scarce [34]. Yet, a temporal perspective is critical in order to assess potential shifts in stable states and tipping points, particularly in dry edge ecosystems highly vulnerable to desertification. In this regard, Anderegg et al. [45] state that climatic stress has surpassed the physiological and ecosystem-level tolerance of woody species growing at the driest edges of their geographic ranges leading to processes.

The main goal of this study is to describe the CHANS dynamics that have shaped forest structure and dynamics during the last century in a Mediterranean forest ecosystem and to examine the forest resilience to key human and natural system transformations. In particular, we explore whether the forestry management strategy developed in Spain since the beginning of the 20th century (i.e., "Ordenaciones de Montes" [46–48]) has been able to bolster the social-ecological forest resilience by reconciling multiple human use demands under intensifying globalization, local socio-demographic transitions and global environmental changes while maintaining key ecosystem function and services in the public "Las Pegueras" forest of Cuellar, Segovia. In previous studies we have shown evidence of resilience in tree growth and stand dynamics in Las Pegueras Forests along the last century and considering teleconnections [49,50]. Here we investigate for the first time resilience of the managed system (sensu Nikinmaa et al. [38]) describing last century patterns in key forest functions and products. To pursue this research goal, we have integrated human and natural system data from historical management records relevant to the case study of Las Pegueras Forest in Central Spain and have interpreted observed patterns following a CHANS conceptual framework. The studied forest is publicly owned and has been managed by the public administration since the beginning of the 20th century.

2. Materials and Methods

2.1. Biophysical Description of Las Pegueras Forest

Las Pegueras Forest (≈7000 ha) is located in Central Spain (41°21′ N, 4°12′ W; Figure 1) and grows on sandy, unconsolidated (inland dunes), siliceous and flat soils at altitudes ranging from 820 to 880 m asl [51]. The annual mean temperature is 12 °C and the annual rainfall is around 460 mm, although the annual rainfall displays high interannual variability [52]. The summer is characterized by severe drought which, together with the low soil water-holding capacity, constrains the establishment of new cohorts [53].

The dominant species is *Pinus pinaster* Ait., commonly known as Maritime pine, representing more than 90% of the trees, while the remaining trees are montane pines (*Pinus sylvestris* L. and *Pinus nigra* Arn.), oaks (*Quercus ilex* L., *Q. faginea* Lam. and *Q. pyrenaica* Willd.) and riparian species (*Salix* spp, *Populus* spp, *Alnus glutinosa* (L.) Gaertn. and *Fraxinus* sp). The understory is composed of disperse annual plants and dwarf shrubs such as *Thymus* spp and *Genista* spp. The above-mentioned site characteristics make the study area the dry edge of *P. pinaster* distribution in its natural range.

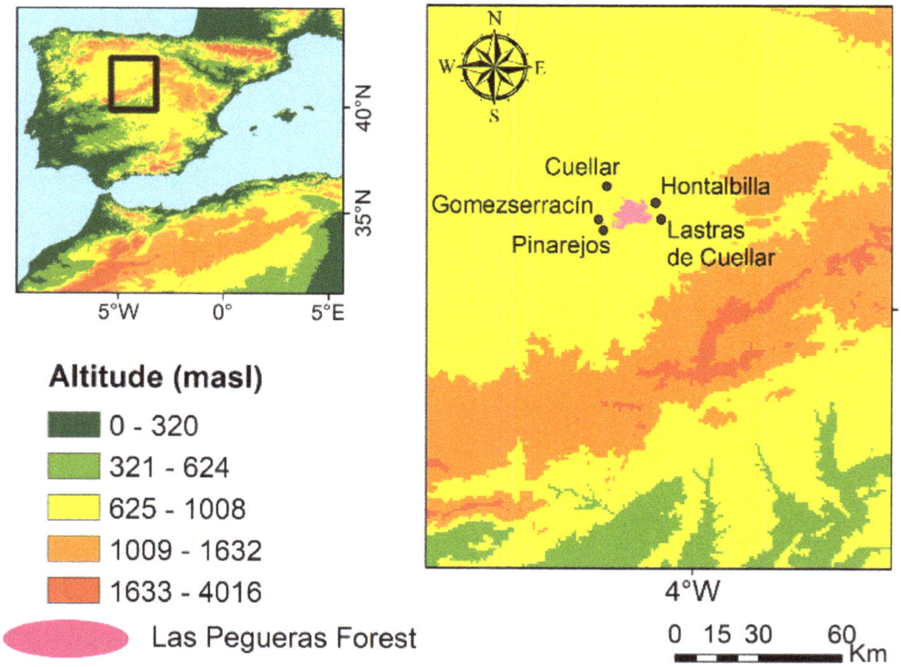

Figure 1. Location of Las Pegueras Forest and the five villages surrounding the forest.

2.2. Socioeconomic Context & Historical Data

Las Pegueras Forest is a public forest belonging to 36 villages although it is managed as a single administrative entity by the autonomous community public forest administration of the Castilla y Leon regional government (Figure 1). The Pegueras forest is considered a "public utility" forest ("Monte de Utilidad Pública") by the Junta de Castilla y Leon regional government. The "public utility" status was accorded during the 19th century as a legal mechanism for protecting socially and ecologically valuable forests, from a preservationist point of view, from extractive economic uses that could degrade their social or ecological protective nature while also preventing their privatization [54,55]. In the case of Las Pegueras forest, its economic use has been, in addition, regulated since 1912 through a forest management plan ("Instrucciones de Ordenación de Montes" [46,47]). This system enables the public administration to guarantee a rational, socially beneficial use of the resources of a given public or private forest. This forest management system in Las Pegueras managed by the Spanish state forest administration first and the autonomous community later has been maintained through changing political conditions in Spain, ranging from the Primo de Rivera and Franco dictatorships, the Spanish Civil War, and finally the democratic regime which initiated the integration of Spain in European Union governance and regional administrative devolution (i.e., from the central Spanish government to autonomous communities). The management plans for Las Pegueras Forest are reviewed every ten years (Table 1). These plans lay down specific spatial and temporal prescriptions for the exploitation and conservation of the forest. The forest (≈7000 ha) is divided into 10 management units which are selectively exploited using the permanent block system in a rotation period of 80 years. In addition, the uniform shelterwood system is applied over a 20-year regeneration period as the method to achieve natural regeneration. When a proper number of new individuals is not achieved, forest managers turn to artificial regeneration. The thinning regime is low. Each management unit is split into four blocks where the trees are expected to be 0–20, 21–40, 41–60, 60–80-year-old.

Table 1. Year of implementation and monitoring period of the management plan and its revisions.

Forest Management Plan	Year	Monitoring Period
Original management plan *	1912	-
1st Revision *	1922	1912–1922
2nd Revision	1932	1922–1932
3rd Revision	1942	1932–1942
4th Revision	1952	1942–1952
5th Revision	1962	1952–1962
6th Revision	1972	1962–1972
7th Revision	1982	1972–1982
8th Revision	1993	1982–1993
9th Revision	2003	1993–2003
10th Revision *	2013	2003–2013

* Data not available.

We obtained the population temporal series of the five villages surrounding the forest (number of inhabitants) from the Spanish Statistical Institute (http://www.ine.es/, accessed on 17 May 2021). These five villages have a close economic relationship with the forest in terms of economic use of ecosystem goods and services. Long-term resin production and imports/export data in Spain and other international markets were obtained from the Spanish Ministry of Agriculture. In order to characterize climatic change impacts in Las Pegueras forest, we calculated the 12-month Standardised Precipitation-Evapotranspiration Index (SPEI_12) for the study area (https://spei.csic.es/spei_database, accessed on 17 May 2021) over the last century [56]. Negative values of SPEI_12 indicate a negative water cumulative balance as a function of precipitation and temperature at a 12-month temporal scale.

2.3. *The Coupled Human and Natural Systems (CHANS) Conceptual Framework*

The CHANS scheme is composed of five major components: (i) sending, receiving, spillover systems; (ii) material, energy or information flow; (iii) agents (e.g., landowners, governments, companies); (iv) economic, political, environmental feedbacks and, (v) environmental and socioeconomic impacts (e.g., loss of biodiversity and ecosystem services, displacement of local people) [57]. CHANS theory has suggested that the compounded interactions of teleconnection and globalization should be integrated into what they define as the telecoupling framework [57]. This approach, which we will be applying in this study, allows for the identification of exchange flows of material/energy or information, agents promoting or hindering these flows, and the environmental and socioeconomic impacts of telecoupling within and between distant CHANS [36].

To cope with the need of considering the temporal complexity of CHANS interactions, Seijo and Gray [58] and Steen-Adams et al. [59] proposed and developed a framework focusing on historical data. This approach provides information on the influence of long-term interactions on current and future human and forest conditions in different historical periods [15,59]. Accordingly, we split the temporal series into three historical periods for the analysis of CHANS interactions. For the sake of simplicity and synthesizing the available information and data we divided historical human forest use into three phases: the Pre-industrial phase (1900s–1960s), the Industrial phase (1960s–1980s) and the Post-industrial phase (1980s–to date) though temporal and spatial uses overlap (eg. industrial phase resin tapping was already present in the late 19th century and preindustrial phase firewood collection and pastoralism lasted until well into the industrial 20th century phase). Figure 2 summarizes in a flow chart our operationalization of the historical telecoupling framework for Las Pegueras.

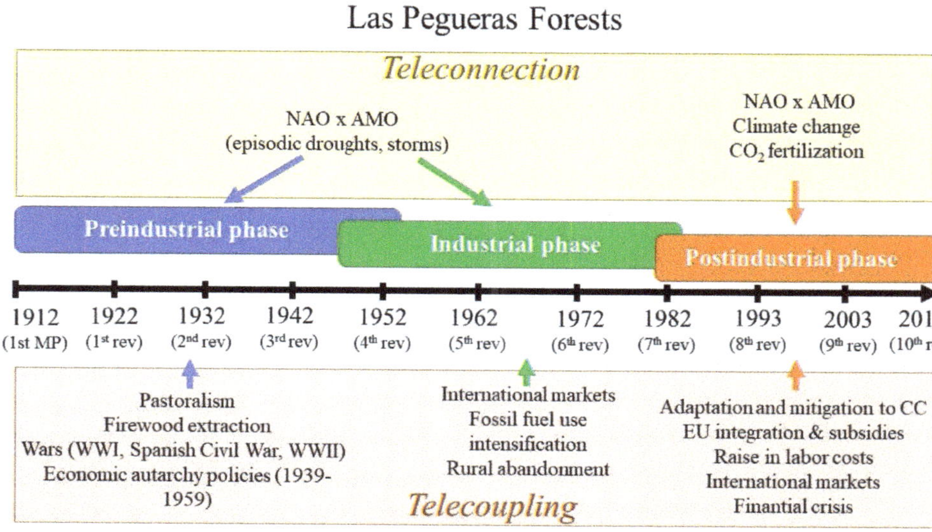

Figure 2. Flow chart of Las Pegueras Forest and the historical telecoupling framework. MP: Management plan. Rev: revision of the management plan.

3. Results and Discussion

Below, we provide a detailed description of the long-term historical data, link these data to telecoupling phenomena during the different historical periods and discuss the effects of the public administration historical management strategies on forest ecosystem resilience.

3.1. Human System Cycles Conditioning the Management Plans of Las Pegueras Forest

3.1.1. From the Preindustrial Phase to the Mid-20th Century

During the Holocene, *P. pinaster* together with *Pinus pinea* L., were the dominant species in these inland dunes [60,61]. However, up until the beginning of the 20th century Las Pegueras Forest, like many other forests in Spain (see for instance Moreno-Fernández et al. [4]) was not regulated by a management plan that ensured the preservation of the forest and the sustainability of its economic uses. The resin sector experienced significant growth during the first decades of the 20th century which led to the implementation of forest management plans to increase the resin tapping potential [62]. The forest was managed for the maximization of resin tapping from the *P. pinaster* trees and, to a lesser extent, timber production. The stand density was low (mean tree density of the blocks is around 390 trees with dbh $\geq$ 20 cm per ha and ranges between 200–800) even in mature stands, to promote trees with large diameter and crown that maximize individual tree resin production [9,63]. Despite the fact the forest was managed for the maximization of resin tapping from the *P. pinaster* trees, the first 1912 management plan prescribed four income-generating economic extractive uses compatible with the forest "public utility" status: (i) resin tapping, (ii) timber production, (iii) firewood extraction from downed branches, litter, twigs as well as other small pieces of wood laying on the ground and (iv) pastoralism with goat herds. Each plan records information on the tree stocks (m^3, number of trees), timber removals (m^3, number of trees, pesetas [Spanish currency before the Euro], €), resin tapping (kg, pesetas, €, number of tapped trees), firewood extraction (pesetas) and pastoral activities (pesetas). Resin tapping takes place annually from March to October in logging units selected by the public administration. Trees were originally tapped using the Hugues method until the middle of the 20th century when it was replaced by a novel system that combines sloped stripes wounds in an upward direction with acid stimulation to promote resin exudation. This method is described in detail in Rodriguez-García et al. [9,63]. Timber extraction was mainly focused on the trees felled during regeneration cuttings of the uniform shelterwood while firewood collection was carried out by the local population. Other functions that have gained importance in later years such as biodiversity conservation, the reduction of soil losses (note that the soils are inland dunes), hunting, fishing, mushroom picking or recreational uses were not contemplated in the management plans.

Once the plan was implemented, the forest yielded higher production of resin, timber stocks and extracted wood (Figure 3). Interestingly, both the resin production and the number of tapped trees peaked during this period (Figure 3A,B). The standing wood volume increased progressively (Figure 3C) while the number of adult trees experienced a rapid increase from 1912 to 1932 when tree density finally stabilized (Figure 3D). This evidence suggests that tree size has increased over the monitoring period.

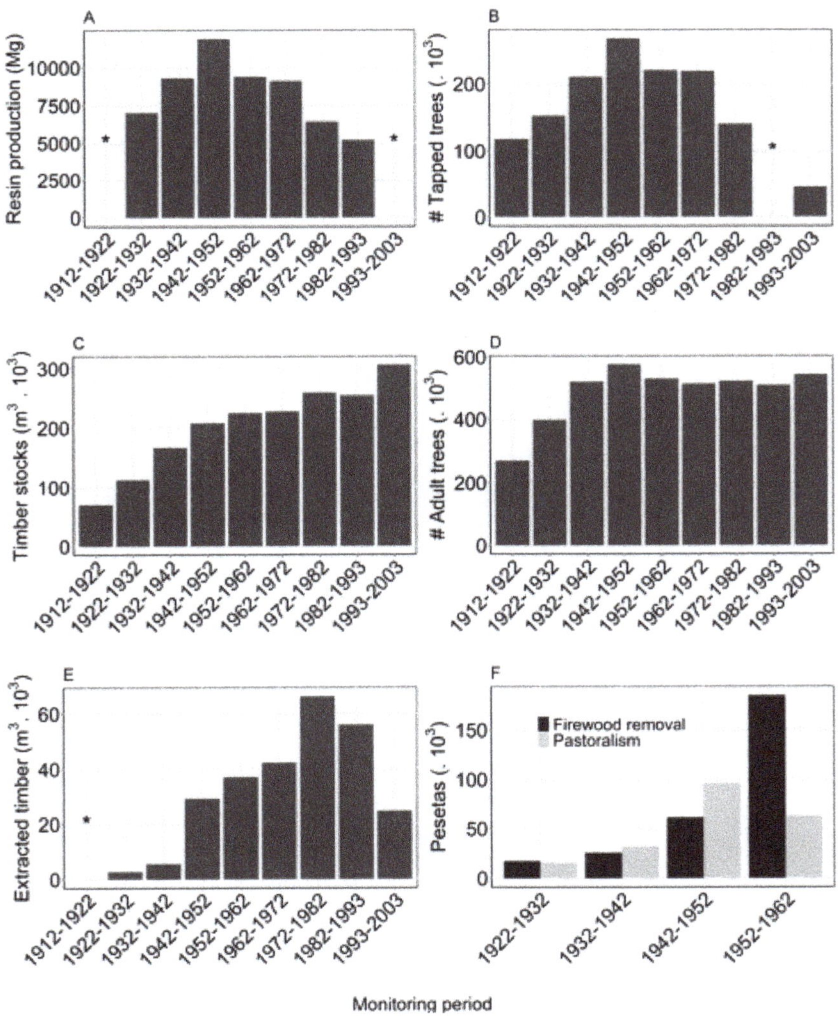

Figure 3. Evolution of resin production (**A**), number of tapped trees (**B**), standing timber stocks (**C**), number of adult trees (dbh > 20 cm) (**D**), extracted timber (**E**), firewood removal and pastoralism (**F**) in Las Pegueras Forest over the study period (1912–2003). Asterisks indicate data non-availability.

During the pre-industrial phase, resin and timber extraction were harmonized with pre-industrial era subsistence economy activities such as hunting, pastoralism and firewood extraction. According to the forest engineer's notes recorded in the management plans, there was an increased social demand for firewood extraction rights. Note that the populations of the villages that own the forest increased during this period (see Figure 4). As a consequence of the increased social demand for firewood, the wood extracted from Las Pegueras increased from 1922 to the mid-20th century (Figure 3F). Similarly, incomes generated from pastoralism reached their maximum values from 1942 to 1952. At this point, it is important to take into account that industrialization in Spain lagged behind other European countries although the economic transition was spatially and temporally very heterogeneous across the Iberian Peninsula and was delayed due to the economic isolation faced by the Franco regime. For example, firewood was still a key energy source in Spain during the first decades of the 20th [64] century and, together with pastoralism, was a fundamental subsistence economy activity for local populations.

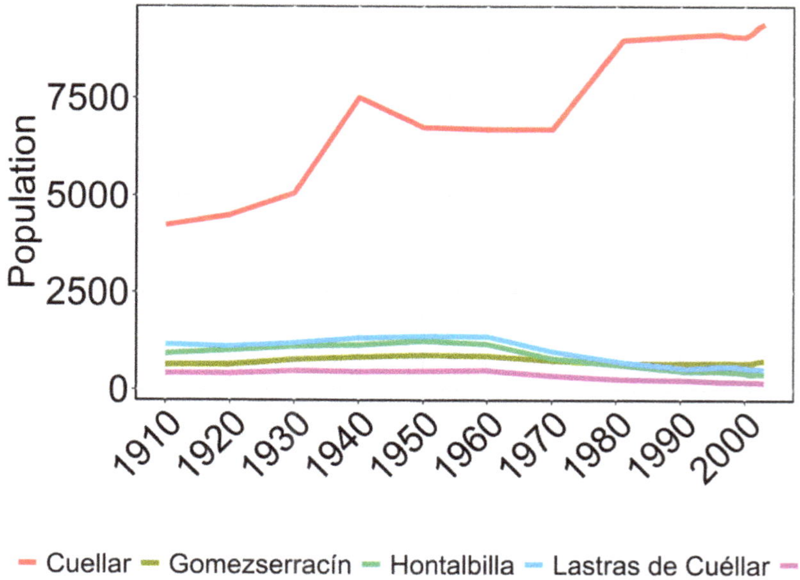

Figure 4. Population trends (number of inhabitants per village) of the five villages that own the forest.

The Las Pegueras Forest, was directly affected by the Spanish Civil War (1936–1939) and indirectly by the First and Second World Wars. The Spanish Civil War resulted in a change in the political regime, that economically isolated Spain from Europe and interrupted the increased integration of Spanish and European markets observed since the end of the 19th century [65]. Additionally, the decades of the 1950s–1960s were especially dry (see low values for SPEI in Figure 5). Despite social-economic upheaval and harsh climatic conditions, timber stocks, resin production and firewood extraction displayed an increment in Las Pegueras Forest.

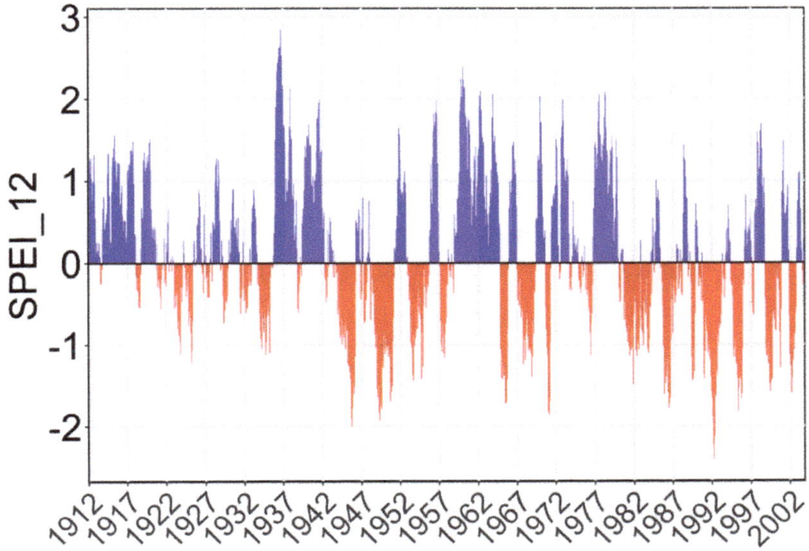

Figure 5. 12-month Standardized Precipitation Evapotranspiration Index (SPEI_12) for the study area from 1902 to 2018.

In sum, during the pre-industrial phase, the public administration management regime was superimposed on traditional ecological knowledge (TEK)-based extractive economic uses by local communities [66,67]. Though the impact on overall forest resilience of these TEK-based practices (firewood extraction, pastoralism) remains an underresearched topic (for instance, the impact on soil nutrient availability due to goat manure) it is clear that the gradual disappearance of these practices led to an increase in standing timber stocks and forest cover based on the evidence available to this study.

3.1.2. From the Mid-20th Century to Industrialization

Spain underwent important economic reforms in the 1960s, which stimulated internal economic demand and jumpstarted the industrialization of the country and the forest sector through a gradual liberalization of the economy [65,68]. This led to an increment in the cost of field labor and progressive rural hollowing or abandonment. Another important consequence of industrialization and the greater interconnectedness of the Spanish with the global economy were the changes in the source of heat energy for homes and industry owing to an increment in the importation of fossil fuels [65,69]. In Las Pegueras Forest, this seems to be related to the observed drop in firewood extraction in the 1960s (Figure 3F). Since the beginning of the 20th century, forest managers recommended phasing out this activity because firewood represents one of the few sources of nutrients in these sandy soils. Finally, the phasing out of pastoralism was also recommended as goats remove shrubs and lichens that contribute to control soil erosion and to avoid damage on the regeneration. Probably, the changes in local energy sourcing, which may have led to a consequent drop in the demand for firewood from local communities, allowed forest managers to restrict firewood extraction and meet this goal. Paradoxically, therefore, as carbon emissions from fossil fuels in the surrounding region of Las Pegueras increased, due to greater telecoupling with international energy markets, local carbon sequestration probably increased as a result of increased forest cover. Again, this possible new function of the forest ecosystem as a carbon sink deserves further research though the issue is beyond the scope of this study.

In this phase both resin variables, i.e., resin production and the number of tapped trees decreased in Las Pegueras Forest (Figure 3A,B, respectively). Similar to the resin tapping trends observed in Las Pegueras, national resin production initially peaked at the beginning of the 1960s industrial boom with productions larger than 50,000 tons and then dropped dramatically [62]. This drop can be linked to the liberalization of the Spanish economy and the emergence of new producing countries such as China and, to a lesser extent, Brazil and Indonesia (see Liu [36] for export trends of forest products from China from the 1960s to 2010s). The incorporation of these new producers in addition to the greater use of fossil fuels by rural communities resulted not only in a telecoupling feedback in Las Pegueras Forest, but also in the entire Iberian Peninsula [70,71] where the emergence of distant resin producing countries due to lower field labor costs resulted in a local reduction of resin tapping in Las Pegueras Forest. Telecoupling feedbacks also led to an increase in the consumption of resin by-products, which resulted in a deficit in their production at the national and also at the European level [71,72]. Consequently, according to the management plans, resin tapping activity in Las Pegueras was maintained thanks to economic subsidies from the public administration.

The drop in resin production in Las Pegueras Forest coincides with a drop in population in the four smallest villages that own the forest (Figure 4). This phenomenon is known as rural abandonment or hollowing. Rural abandonment was common in Spain during this period but was somewhat more contained in resin-producing villages as resin tapping served to generate rural employment [71]. However, it became increasingly difficult to recruit workers for resin tapping due to the arduous nature of the job with scarce rural workers preferring more stable jobs in the industrial belt surrounding Cuéllar. Cuéllar, with more than 5000 inhabitants, in fact did not seem to be affected by demographic abandonment which may have been driven, in part, by the contraction of the resin sector.

Similar to the observed trends for resin variables, the extracted timber followed a bell-shaped reaching a maximum during the 1970s.

3.1.3. From the Late 20th Century to the Post-Industrial Phase

The aforementioned resin sector telecoupling feedback started changing at the beginning of the 21th century when China experienced an industrialization process similar to the one that had occurred in Spain during the 1950s and 1960s. Both processes have resulted in similar consequences, such as rural abandonment, rural aging and raised field-labor costs [73,74]. European Union integration together with the new paradigm in European rural governance via the European budget (tiny rural repopulation and recovery of traditional uses) opened up by post-2008 financial crisis European stimuli spending led paradoxically to a re-emergence of the demand for resin tapping. However, this increased resin tapping activity has been accomplished through public administration subsidies which may undermine its economic sustainability in the context of greater public sector indebtedness. In this regard, Soliño et al. [75] analyzed local community perception of subsidized resin-tapped *P. pinaster* forests in Central Spain. They highlighted that, overall, local communities agreed with the idea of promoting resin tapping activities again in these forests and that its promotion could result in lower wildfire risk due to associated understory fuel reduction (for instance, a resin tapper needs to tap about 1000 trees to obtain profitability which requires an open forest structure to facilitate access to individual trees) and the maintenance of rural population. Again this is a topic that needs to be further researched from the point of view of overall CHANS sustainability and the emerging discussion of payment for ecosystem services schemes at a European and Spanish level [76].

3.2. *Ecological Legacy Effects of Past Uses and Emerging Resilience Challenges*

Our results indicate an increase of the standing timber stocks together with an increment of productivity in terms of resin production and timber extraction since the implementation of the first management plan. Though many individual trees have been damaged by resin tapping with the consequent loss of radial growth [77], standing timber stocks were four times larger in 2003 than in 1912. This huge increment in forest cover is positively linked to other ecosystem functions. First, the increase in forest cover is expected to reduce wind erosion and promote soil consolidation and stabilization [78,79]. Secondly, the maximization of standing stocks favors the fixation of carbon and contributes to the mitigation of climate change [4,80]. As discussed previously, the decline in the two main income-generating forest functions, resin tapping and timber removal, were influenced by telecoupling feedbacks over which the historical management plans had little influence rather than by a dramatic fall in the potential productivity of these activities. In addition to changing telecoupling feedbacks the evolution of this CHANS is largely driven by the impacts of teleconnection with the global climate system which might threaten social-ecological resilience. The maintenance of wood stocks over the study period, however, indicates that ecological resilience has not been decreased due to varying teleconnection feedbacks (i.e., climate change). The extracted timber and resin production —that is, human use—has declined due to telecoupling feedbacks, but they have the potential to be restored or maximized with better human system management.

Madrigal-González et al. [50] have studied global climate impacts via transformations in the North Atlantic Oscillation (NAO) and the Atlantic Mediterranean Oscillation (AMO) on tree dynamics in Las Pegueras Forest (engineering resilience). For instance, the NAO affects tree growth in Las Pegueras through increased aridity due to reduced annual precipitation. Hence, the frequency of dry years has increased since the 1980s (Figure 5). Similarly but at the forest level (ecological resilience), Madrigal-González et al. [49] studied the demographic resilience of Las Pegueras over the last century and identified that the ecological resilience is high but the forecasted increment of droughts episodes, in terms of severity and frequency, could exceed the resilience of the system. In this regard, thinning operations prescribed by the management plans (applied from the beginning of the rotation period to the beginning of the regeneration period) attenuated the impact of dry years on tree growth [81]. These cycles are not, however, explicitly accounted for in management prescriptions plans which are based on average climatic expectations.

Regeneration is a key process for ensuring the renewal and persistence of forests [82]. Success in the regeneration of Mediterranean tree species is constrained by summer droughts, among other factors [83], which are forecasted to increase in severity, extent and frequency [84–86]. In our case study, the uniform shelterwood cuttings, which are carried out over the last 20 years of the rotation period, achieved the establishment of new cohorts and, therefore, the renewal of the forest [87]. Although *P. pinaster* is a light demanding species, summer climatic conditions can be so harsh that partial cover is needed to ensure regeneration. Then, to alleviate the effects of the limiting summer conditions on seedling establishment and to assure sufficient natural regeneration [83], the recent revisions of the management plans extended the regeneration period to 25 years. The uniform shelterwood method manipulates stand density to optimize seedling survival along the shade-drought tolerant tradeoff that is key in the assembly of Mediterranean forests [88,89]. Given long term forest function maintenance the management method (permanent blocks and the uniform shelterwood system) used in Las Pegueras Forest seems to have successfully contributed to CHANS resilience though more research is needed to evaluate its impact on rural depopulation. This forest, however, faces new challenges mainly associated with climate change adaptation, mitigation and renewable energy production policies in Europe and Spain. Many forests of *P. pinaster* surrounding the study area are affected by decline processes, that are expressed through attenuated forest growth rates, increments in mortality and failure in regeneration [53,90]. This situation is aggravated by the overexploitation of aquifers for agricultural purposes, which results in strong fluctuations of the water table [53,83]. Moreover, advances in stimulants for resin tapping may promote the yield as well as this rural activity [91] though its reformulation as a subsidized payment for ecosystem services scheme may be necessary due to its low profitability and high volatility given international market conditions. Additionally, preliminary studies from an ongoing research project indicate that the tapped *P. pinaster* trees are valid for structural uses [92].

4. Conclusions

The observed long-term ecological forest functionally is intrinsically coupled with global human system telecoupled components and biotic legacy effects resulting from past uses. This study highlights the role of the historical forest management legacies as well as telecoupling phenomena driven by accelerating globalization and climate system teleconnections with current forest dynamics, processes and functionality. The Las Pegueras CHANS has been affected by a civil war, changes in political regimes, telecoupling processes in the global economy as well as drought periods but the persistence and ecological functions provided in part by the historical management regime has successfully mitigated many of these pressures. Therefore, at the forest landscape level, the application of the permanent block method in conjunction with the regeneration stimulated by the uniform shelterwood cuttings system has alleviated the impacts of socioeconomic and climatic changes promoting increased forest resilience. Therefore, our results highlight the social-ecological resilience of the forest which confirms that found at lower hierarchical levels (see Madrigal-González et al. [50] for engineering resilience-tree level and Madrigal-González et al. [49] for ecological resilience–forest level).

The generation of a continued income stream for local populations may have led to stakeholder general acceptance of the governance system but more research into local community perceptions of the governance system is needed. Finally, our results suggest that it is necessary to further research historical legacy effects of past human uses as well as teleconnection and telecoupling feedbacks when analyzing forest ecosystem functions in addition to addressing local and short-term temporal features. Integration of these levels of complexity—local and global feedbacks—may not necessarily increase model complexity but could result in a more realistic description of this ecosystem than models based on ecological processes alone.

Author Contributions: Conceptualization, M.A.Z. and F.S.; methodology, F.S.; formal analysis, D.M.-F.; data curation, D.M.-F.; writing—original draft preparation, D.M.-F.; resources, J.M.-G.; writing—review and editing, F.S., M.A.Z. and J.M.-G.; visualization, D.M.-F.; funding acquisition, M.A.Z. All authors have read and agreed to the published version of the manuscript.

Funding: This research was funded by the Spanish Ministry of Science, Innovation and Universities, grant number RTI2018-096884-B-C32 (DARE). DMF is supported by the Spanish Ministry of Science, Innovation and Universities through the Juan de la Cierva Formación post-doctoral program (FJC2018-037870-I).

Data Availability Statement: The data presented in this study are available on request from the corresponding author.

Acknowledgments: Authors thank forest managers of Junta de Castilla y León and in particular to María Bragado and José Ignacio Quintanilla (ServicioTerritorial de Medioambiente de Segovia, Spain) for their support with the historical archives, interesting discussions and for sharing their expertise with us. We thank the editor Juan Fernández-Manjarrés for his invitation to participate in this special issue and for critical advice along with its conception. We also the two anonymous reviewers for their comments, which helped to improve the manuscript.

Conflicts of Interest: The authors declare no conflict of interest.

References

1. Soga, M.; Gaston, K.J. Extinction of experience: The loss of human-nature interactions. *Front. Ecol. Environ.* **2016**, *14*, 94–101. [CrossRef]
2. Ellis, E.C.; Gauthier, N.; Klein Goldewijk, K.; Bliege Bird, R.; Boivin, N.; Díaz, S.; Fuller, D.Q.; Gill, J.L.; Kaplan, J.O.; Kingston, N.; et al. People have shaped most of terrestrial nature for at least 12,000 years. *Proc. Natl. Acad. Sci. USA* **2021**, *118*, e2023483118. [CrossRef]
3. Bottalico, F.; Pesola, L.; Vizzarri, M.; Antonello, L.; Barbati, A.; Chirici, G.; Corona, P.; Cullotta, S.; Garfì, V.; Giannico, V.; et al. Modeling the influence of alternative forest management scenarios on wood production and carbon storage: A case study in the Mediterranean region. *Environ. Res.* **2016**, *144*, 72–87. [CrossRef]

4. Moreno-Fernández, D.; Díaz-Pinés, E.; Barbeito, I.; Sánchez-González, M.; Montes, F.; Rubio, A.; Cañellas, I. Temporal carbon dynamics over the rotation period of two alternative management systems in Mediterranean mountain Scots pine forests. *For. Ecol. Manage.* **2015**, *348*, 186–195. [CrossRef]
5. Basnou, C.; Vicente, P.; Espelta, J.M.; Pino, J. Of niche differentiation, dispersal ability and historical legacies: What drives woody community assembly in recent Mediterranean forests? *Oikos* **2016**, *125*, 107–116. [CrossRef]
6. Sgroi, F.; Foderà, M.; Dana, L.P.; Mangiapane, G.; Tudisca, S.; Di Trapani, A.M.; Testa, R. Evaluation of payment for ecosystem services in Mediterranean forest: An empirical survey. *Ecol. Eng.* **2016**, *90*, 399–404. [CrossRef]
7. López-Sánchez, A.; Dirzo, R.; Roig, S. Changes in livestock footprint and tree layer coverage in Mediterranean dehesas: A six-decade study based on remote sensing. *Int. J. Remote Sens.* **2017**, 1–17. [CrossRef]
8. Martínez-Jauregui, M.; Herruzo, A.C.; Campos, P.; Soliño, M. Shedding light on the self-consumption value of recreational hunting in European Mediterranean forests. *For. Policy Econ.* **2016**, *69*, 83–89. [CrossRef]
9. Rodríguez-García, A.; Martín, J.A.; López, R.; Mutke, S.; Pinillos, F.; Gil, L. Influence of climate variables on resin yield and secretory structures in tapped *Pinus pinaster* Ait. in central Spain. *Agric. For. Meteorol.* **2015**, *202*, 83–93. [CrossRef]
10. Von Essen, M.; Do Rosário, I.T.; Santos-Reis, M.; Nicholas, K.A. Valuing and mapping cork and carbon across land use scenarios in a Portuguese montado landscape. *PLoS ONE* **2019**, *14*, e0212174. [CrossRef]
11. Calama, R.; Montero, G. Cone and seed production from stone pine (*Pinus pinea* L.) stands in Central Range (Spain). *Eur. J. For. Res.* **2007**, *126*, 23–35. [CrossRef]
12. Morán-Ordóñez, A.; Ameztegui, A.; De Cáceres, M.; de-Miguel, S.; Lefèvre, F.; Brotons, L.; Coll, L. Future trade-offs and synergies among ecosystem services in Mediterranean forests under global change scenarios. *Ecosyst. Serv.* **2020**, *45*, 101174. [CrossRef]
13. Seijo, F.; Godoy, M.M.; Guglielmin, D.; Ciampoli, C.; Ebright, S.; Picco, O.; Defossé, G. Conflicting Frames about Ownership and Land Use Drive Wildfire Ignitions in a Protected Conservation Area. *Environ. Manag.* **2020**, *65*, 448–462. [CrossRef] [PubMed]
14. Muhar, A.; Raymond, C.M.; van den Born, R.J.G.; Bauer, N.; Böck, K.; Braito, M.; Buijs, A.; Flint, C.; de Groot, W.T.; Ives, C.D.; et al. A model integrating social-cultural concepts of nature into frameworks of interaction between social and natural systems. *J. Environ. Plan. Manag.* **2018**, *61*, 756–777. [CrossRef]
15. Steen-Adams, M.M.; Charnley, S.; Adams, M.D. Historical perspective on the influence of wildfire policy, law, and informal institutions on management and forest resilience in a multiownership, frequent-fire, coupled human and natural system in Oregon, USA. *Ecol. Soc.* **2017**, *22*. [CrossRef]
16. Liu, J.; Dietz, T.; Carpenter, S.R.; Folke, C.; Alberti, M.; Redman, C.L.; Schneider, S.H.; Ostrom, E.; Pell, A.N.; Lubchenco, J.; et al. Coupled human and natural systems. *Ambio* **2007**, *36*, 639–649. [CrossRef]
17. Alberti, M.; Asbjornsen, H.; Baker, L.A.; Brozović, N.; Drinkwater, L.E.; Brozovic, N.; Drzyzga, S.A.; Jantz, C.A.; Fragoso, J.; Holland, D.S.; et al. Research on Coupled Human and Natural Systems (CHANS): Approach, Challenges, and Strategies. *Bull. Ecol. Soc. Am.* **2011**, *92*, 218–228. [CrossRef]
18. Müller, F. Indicating ecosystem and landscape organisation. *Ecol. Indic.* **2005**, *5*, 280–294. [CrossRef]
19. Burkhard, B.; Fath, B.D.; Müller, F. Adapting the adaptive cycle: Hypotheses on the development of ecosystem properties and services. *Ecol. Modell.* **2011**, *222*, 2878–2890. [CrossRef]
20. Crous-Duran, J.; Graves, A.R.; García de Jalón, S.; Kay, S.; Tomé, M.; Burgess, P.J.; Giannitsopoulos, M.; Palma, J.H.N. Quantifying Regulating Ecosystem Services with Increased Tree Densities on European Farmland. *Sustainability* **2020**, *12*, 6676. [CrossRef]
21. Duncker, P.S.; Raulund-Rasmussen, K.; Gundersen, P.; Katzensteiner, K.; De Jong, J.; Ravn, H.P.; Smith, M.; Eckmüllner, O.; Spiecker, H. How Forest Management affects Ecosystem Services, including Timber Production and Economic Return: Synergies and Trade-Offs. *Ecol. Soc.* **2012**, *17*, art50. [CrossRef]
22. Pukkala, T. Which type of forest management provides most ecosystem services? *For. Ecosyst.* **2016**, *3*, 1–16. [CrossRef]
23. Kim, G.S.; Lim, C.-H.; Kim, S.J.; Lee, J.; Son, Y.; Lee, W.-K. Effect of National-Scale Afforestation on Forest Water Supply and Soil Loss in South Korea, 1971–2010. *Sustainability* **2017**, *9*, 1017. [CrossRef]
24. Tu, H.M.; Chen, H.M. From deforestation to afforestation: Effect of slopeland use policies on land use/cover change in Taiwan. *Land Use Policy* **2020**, *99*, 105038. [CrossRef]
25. Zomer, R.J.; Trabucco, A.; Bossio, D.A.; Verchot, L.V. Climate change mitigation: A spatial analysis of global land suitability for clean development mechanism afforestation and reforestation. *Agric. Ecosyst. Environ.* **2008**, *126*, 67–80. [CrossRef]
26. Nijnik, M.; Halder, P. Afforestation and reforestation projects in South and South-East Asia under the Clean Development Mechanism: Trends and development opportunities. *Land Use Policy* **2013**, *31*, 504–515. [CrossRef]
27. Hansen, W.D.; Braziunas, K.H.; Rammer, W.; Seidl, R.; Turner, M.G. It takes a few to tango: Changing climate and fire regimes can cause regeneration failure of two subalpine conifers. *Ecology* **2018**, *99*, 966–977. [CrossRef]
28. Davis, K.T.; Dobrowski, S.Z.; Higuera, P.E.; Holden, Z.A.; Veblen, T.T.; Rother, M.T.; Parks, S.A.; Sala, A.; Maneta, M.P. Wildfires and climate change push low-elevation forests across a critical climate threshold for tree regeneration. *Proc. Natl. Acad. Sci. USA* **2019**, *116*, 6193–6198. [CrossRef]
29. Calama, R.; Gordo, J.; Mutke, S.; Conde, M.; Madrigal, G.; Garriga, E.; Arias, M.J.; Piqué, M.; Gandía, R.; Montero, G.; et al. Decline in commercial pine nut and kernel yield in Mediterranean stone pine (*Pinus pinea* L.) in Spain Rafael. *iForest* **2020**, *13*, 251–260. [CrossRef]
30. Coomes, D.A.; Flores, O.; Holdaway, R.; Jucker, T.; Lines, E.R.; Vanderwel, M.C. Wood production response to climate change will depend critically on forest composition and structure. *Glob. Chang. Biol.* **2014**, *20*, 3632–3645. [CrossRef]

31. Glantz, M.; Katz, R.; Nicholls, N. *Teleconnections Linking Worldwide Climate Anomalies*; Cambridge Univ. Press: Cambridge, UK, 1991.
32. Kastner, T.; Erb, K.H.; Nonhebel, S. International wood trade and forest change: A global analysis. *Glob. Environ. Chang.* **2011**, *21*, 947–956. [CrossRef]
33. Nyström, M.; Jouffray, J.B.; Norström, A.V.; Crona, B.; Søgaard Jørgensen, P.; Carpenter, S.R.; Bodin Galaz, V.; Folke, C. Anatomy and resilience of the global production ecosystem. *Nature* **2019**, *575*, 98–108. [CrossRef] [PubMed]
34. Martínez-Valderrama, J.; Sanjuán, M.E.; del Barrio, G.; Guirado, E.; Ruiz, A.; Maestre, F.T. Mediterranean landscape re-greening at the expense of South American forests. *Land* **2021**, *10*, 204. [CrossRef]
35. Urbieta, I.R.; Zavala, M.A.; Marañón, T. Human and non-human determinants of forest composition in southern Spain: Evidence of shifts towards cork oak dominance as a result of management over the past century. *J. Biogeogr.* **2008**, *35*, 1688–1700. [CrossRef]
36. Liu, J. Forest Sustainability in China and Implications for a Telecoupled World. *Asia Pacific Policy Stud.* **2014**, *1*, 230–250. [CrossRef]
37. Bone, C.; Moseley, C.; Vinyeta, K.; Bixler, R.P. Employing resilience in the United States Forest Service. *Land Use Policy* **2016**, *52*, 430–438. [CrossRef]
38. Nikinmaa, L.; Lindner, M.; Cantarello, E.; Jump, A.S.; Seidl, R.; Winkel, G.; Muys, B. Reviewing the Use of Resilience Concepts in Forest Sciences. *Curr. For. Reports* **2020**, *6*, 61–80. [CrossRef]
39. Pimm, S.L. The complexity and stability of ecosystems. *Nature* **1984**, *307*, 321–326. [CrossRef]
40. Holling, C.S. Resilience and Stability of Ecological Systems. *Annu. Rev. Ecol. Syst.* **2003**, 245–256. [CrossRef]
41. Englund, O.; Berndes, G.; Cederberg, C. How to analyse ecosystem services in landscapes—A systematic review. *Ecol. Indic.* **2017**, *73*, 492–504. [CrossRef]
42. Schulp; Burkhard, B.; Maes, J.; Van Vliet, J.; Verburg, P.H. Uncertainties in ecosystem service maps: A comparison on the European scale. *PLoS ONE* **2014**, *9*. [CrossRef] [PubMed]
43. Cruz-Alonso, V.; Ruiz-Benito, P.; Villar-Salvador, P.; Rey-Benayas, J.M. Long-term recovery of multifunctionality in Mediterranean forests depends on restoration strategy and forest type. *J. Appl. Ecol.* **2019**, *56*, 745–757. [CrossRef]
44. Sutherland, I.J.; Bennett, E.M.; Gergel, S.E. Recovery trends for multiple ecosystem services reveal non-linear responses and long-term tradeoffs from temperate forest harvesting. *For. Ecol. Manage.* **2016**, *374*, 61–70. [CrossRef]
45. Anderegg, W.R.L.; Anderegg, L.D.L.; Kerr, K.L.; Trugman, A.T. Widespread drought-induced tree mortality at dry range edges indicates that climate stress exceeds species' compensating mechanisms. *Glob. Chang. Biol.* **2019**, *25*, 3793–3802. [CrossRef] [PubMed]
46. Mackay, E. *Fundamentos y Métodos de la Ordenación de Montes. Primera Parte*; Escuela Especial de Ingenieros de Montes: Madrid, Spain, 1944.
47. Mackay, E. *Fundamentos y Métodos de la Ordenación de Montes. Segunda Parte*; Escuela Especial de Ingenieros de Montes: Madrid, Spain, 1949.
48. Olazabal, L. *Ordenación y Valoración de Montes*; Imprenta de Moreno y Rojas: Madrid, Spain, 1883.
49. Madrigal-González, J.; Herrero, A.; Ruiz-Benito, P.; Zavala, M.A. Resilience to drought in a dry forest: Insights from demographic rates. *For. Ecol. Manage.* **2017**, *389*, 167–175. [CrossRef]
50. Madrigal-González, J.; Andivia, E.; Zavala, M.A.; Stoffel, M.; Calatayud, J.; Sánchez-Salguero, R.; Ballesteros-Cánovas, J. Disentangling the relative role of climate change on tree growth in an extreme Mediterranean environment. *Sci. Total Environ.* **2018**, *642*, 619–628. [CrossRef]
51. Gómez-Sanz, V.; García-Viñas, J.I. Soil moisture spatio-temporal behavior of Pinus pinaster stands on sandy flatlands of central Spain. *For. Syst.* **2011**, *20*, 293–302. [CrossRef]
52. Madrigal-González, J.; Hantson, S.; Yue, C.; Poulter, B.; Ciais, P.; Zavala, M.A. Long-term Wood Production in Water-Limited Forests: Evaluating Potential CO2 Fertilization Along with Historical Confounding Factors. *Ecosystems* **2015**, *18*, 1043–1055. [CrossRef]
53. Moreno-Fernández, D.; Montes, F.; Sánchez-González, M.; Gordo, F.J.; Cañellas, I. Regeneration dynamics of mixed stands of *Pinus pinaster* Ait. and *Pinus pinea* L. in Central Spain. *Eur. J. For. Res.* **2018**, *137*, 17–27. [CrossRef]
54. Bauer, E. *Los Montes de España en la Historia*; Ministerio de Agricultura y Escuela Técnica Superior de Ingenieros de Montes: Madrid, Spain, 1991.
55. Pérez-Soba, I. Los montes de utilidad pública: Un patrimonio con mucho pasado y mucho futuro. *Ambienta* **2013**, *104*, 54–72. (In Spanish)
56. Vicente-Serrano, S.M.; Beguería, S.; López-Moreno, J.I. A multiscalar drought index sensitive to global warming: The standardized precipitation evapotranspiration index. *J. Clim.* **2010**, *23*, 1696–1718. [CrossRef]
57. Liu, J.; Hull, V.; Batistella, M.; Defries, R.; Dietz, T.; Fu, F.; Hertel, T.W.; Izaurralde, R.C.; Lambin, E.F.; Li, S.; et al. Framing Sustainability in a Telecoupled World. *Ecol. Soc.* **2013**, *18*, 26. [CrossRef]
58. Seijo, F.; Gray, R. Pre-Industrial Anthropogenic Fire Regimes in Transition: The Case of Spain and its Implications for Fire Governance in Mediterranean Type Biomes. *Soc. Hum. Ecol.* **2012**, *19*, 58–59.
59. Steen-Adams, M.M.; Langston, N.; Adams, M.D.O.; Mladenoff, D.J. Historical framework to explain long-term coupled human and natural system feedbacks: Application to a multiple-ownership forest landscape in the Northern Great Lakes region, USA. *Ecol. Soc.* **2015**, *20*, 28. [CrossRef]

60. Franco-Múgica, F.; García-Antón, M.; Maldonado-Ruiz, J.; Morla-Juaristi, C.; Sainz-Ollero, H. Ancient pine forest on inland dunes in the Spanish northern meseta. *Quat. Res.* **2005**, *63*, 1–14. [CrossRef]
61. Morales-Molino, C.; Postigo-Mijarra, J.M.; Morla, C.; García-Antón, M. Long-term persistence of Mediterranean pine forests in the Duero Basin (central Spain) during the Holocene: The case of *Pinus pinaster* Aiton. *Holocene* **2012**, *22*, 561–570. [CrossRef]
62. Picardo, A.; Pinillos, F. La resinación en España y en el mundo en 2013: Situación y perspectivas. In Proceedings of the II International Symposium on Natural Resins; Ministerio de Agricultura, Alimentación y Medio Ambiente: Madrid, Spain, 2013; pp. 30–53.
63. Rodríguez-García, A.; López, R.; Martín, J.A.; Pinillos, F.; Gil, L. Resin yield in Pinus pinaster is related to tree dendrometry, stand density and tapping-induced systemic changes in xylem anatomy. *For. Ecol. Manage.* **2014**, *313*, 47–54. [CrossRef]
64. Iriarte-Goñi, I.; Infante-Amate, J. Continuity, change, and geographical differences in Spain's firewood consumption: A new estimation (1860-2010). *Hist. Agrar.* **2019**, 33–57. [CrossRef]
65. Infante-Amate, J.; Soto, D.; Aguilera, E.; García-Ruiz, R.; Guzmán, G.; Cid, A.; De Molina, M.G. The Spanish transition to industrial metabolism: Long-term material flow analysis (1860–2010). *J. Ind. Ecol.* **2015**, *19*, 866–876. [CrossRef]
66. Berkes, F.; Colding, J.; Folke, C. Rediscovery of traditional ecological knowledge as adaptive management. *Ecol. Appl.* **2000**, *10*, 1251–1262. [CrossRef]
67. Hosen, N.; Nakamura, H.; Hamzah, A. Adaptation to climate change: Does traditional ecological knowledge hold the key? *Sustainability* **2020**, *12*, 676. [CrossRef]
68. Molero, J. Industrialization and internationalization in the Spanish economy. In *The Emerging Industrial Structure of the Wider Europe*; McGowan, F., Radosevic, S., von Tunzelman, N., Eds.; Taylor & Francis: Oxford, UK, 2004.
69. Iriarte-Goñi, I. Forests, fuelwood, pulpwood, and lumber in Spain, 1860-2000: A non-declensionist story. *Environ. Hist. Durh. N. C.* **2013**, *18*, 333–359. [CrossRef]
70. Palma, A.; Pereira, J.M.; Soares, P. Resin tapping activity as a contribution to the management of maritime pine forest. *For. Syst.* **2016**, *25*. [CrossRef]
71. Pérez Ortuño, S.; Robredo García, F.; Tellez Ayuga, E.; Belda Fullana, C. Effects of the crisis in the resin sector on the demography of rural municipalities in spain. *For. Syst.* **2013**, *22*, 39–46. [CrossRef]
72. Allué, M.; Picardo, A.; Pinillos, F. *Diagnóstico y Propuestas de Actuación en el Sector Resinero*; CESEFOR: Soria, Spain, 2004.
73. Xie, H.; Wang, P.; Yao, G. Exploring the Dynamic Mechanisms of Farmland Abandonment Based on a Spatially Explicit Economic Model for Environmental Sustainability: A Case Study in Jiangxi Province, China. *Sustainability* **2014**, *6*, 1260–1282. [CrossRef]
74. Liao, L.; Long, H.; Gao, X.; Ma, E. Effects of land use transitions and rural aging on agricultural production in China's farming area: A perspective from changing labor employing quantity in the planting industry. *Land Use Policy* **2019**, *88*, 104152. [CrossRef]
75. Soliño, M.; Yu, T.; Alía, R.; Auñón, F.; Bravo-Oviedo, A.; Chambel, M.R.; de Miguel, J.; del Río, M.; Justes, A.; Martínez-Jauregui, M.; et al. Resin-tapped pine forests in Spain: Ecological diversity and economic valuation. *Sci. Total Environ.* **2018**, *625*, 1146–1155. [CrossRef]
76. Zingari, P.C.; Del Gobbo, G. Ecosystem services, European Union policies, and stakeholders' participation. In *Ecosystem Services of Headwater Catchments*; Křeček, J., Haigh, M., Hofer, T., Kubin, E., Promper, C., Eds.; Springer International Publishing: Berlin/Heidelberg, Germany, 2017; pp. 225–238. ISBN 9783319579467.
77. Génova, M.; Caminero, L.; Dochao, J. Resin tapping in *Pinus pinaster*: Effects on growth and response function to climate. *Eur. J. For. Res.* **2014**, *133*, 323–333. [CrossRef]
78. Mosavi, A.; Sajedi-Hosseini, F.; Choubin, B.; Taromideh, F.; Rahi, G.; Dineva, A. Susceptibility Mapping of Soil Water Erosion Using Machine Learning Models. *Water* **2020**, *12*, 1995. [CrossRef]
79. Panagos, P.; Borrelli, P.; Meusburger, K.; Alewell, C.; Lugato, E.; Montanarella, L. Estimating the soil erosion cover-management factor at the European scale. *Land Use Policy* **2015**, *48*, 38–50. [CrossRef]
80. Ontl, T.A.; Janowiak, M.K.; Swanston, C.W.; Daley, J.; Handler, S.; Cornett, M.; Hagenbuch, S.; Handrick, C.; McCarthy, L.; Patch, N. Forest management for carbon sequestration and climate adaptation. *J. For.* **2020**, *118*, 86–101. [CrossRef]
81. Madrigal-González, J.; Zavala, M.A. Competition and tree age modulated last century pine growth responses to high frequency of dry years in a water limited forest ecosystem. *Agric. For. Meteorol.* **2014**, *192–193*, 18–26. [CrossRef]
82. Lucas-Borja, M.E. Climate change and forest natural regeneration in Mediterranean mountain areas. *For. Res.* **2014**, *3*, 2–3. [CrossRef]
83. Calama, R.; Manso, R.; Lucas-Borja, M.E.; Espelta, J.M.; Piqué, M.; Bravo, F.; Del Peso, C.; Pardos, M. Natural regeneration in Iberian pines: A review of dynamic processes and proposals for management. *For. Syst.* **2017**, *26*, eR02S. [CrossRef]
84. Giorgi, F.; Lionello, P. Climate change projections for the Mediterranean region. *Glob. Planet. Change* **2008**, *63*, 90–104. [CrossRef]
85. Gao, X.; Giorgi, F. Increased aridity in the Mediterranean region under greenhouse gas forcing estimated from high resolution simulations with a regional climate model. *Glob. Planet. Change* **2008**, *62*, 195–209. [CrossRef]
86. Turco, M.; Ceglar, A.; Prodhomme, C.; Soret, A.; Toreti, A.; Francisco, J.D.-R. Summer drought predictability over Europe: Empirical versus dynamical forecasts. *Environ. Res. Lett.* **2017**, *12*, 84006. [CrossRef]
87. Andivia, E.; Madrigal-González, J.; Villar-Salvador, P.; Zavala, M.A. Do adult trees increase conspecific juvenile resilience to recurrent droughts? Implications for forest regeneration. *Ecosphere* **2018**, *9*, e02282. [CrossRef]
88. Zavala, M.A.; Espelta, J.M.; Retana, J. Constraints and trade-offs in Mediterranean plant communities: The case of holm oak-Aleppo pine forests. *Bot. Rev.* **2000**, *66*, 119–149. [CrossRef]

89. Sánchez-Gómez, D.; Valladares, F.; Zavala, M.A. Performance of seedlings of Mediterranean woody species under experimental gradients of irradiance and water availability: Trade-offs and evidence for niche differentiation. *New Phytol.* **2006**, *170*, 795–806. [CrossRef]
90. Prieto-Recio, C.; Martín-García, J.; Bravo, F.; Diez, J.J. Unravelling the associations between climate, soil properties and forest management in *Pinus pinaster* decline in the Iberian Peninsula. *For. Ecol. Manage.* **2015**, *356*, 74–83. [CrossRef]
91. Michavila, S.; Rodríguez-García, A.; Rubio, F.; Gil, L.; López, R. Salicylic and citric acid as promising new stimulants for resin tapping in maritime pine (*Pinus pinaster* Ait.). *For. Syst.* **2020**, *29*, 07. [CrossRef]
92. SustForests El Proyecto SustForest Plus Valida la Producción de Madera Resinada Como uso Estructural. Available online: https://www.sust-forest.eu/es/noticia/el-proyecto-sustforest-plus-valida-la-produccion-de-madera-resinada-para-uso-estructurall (accessed on 10 August 2021).

forests

MDPI

Article

Winners and Losers in Energy Transition: Study Case of Wood Biomass Power-Plants Implementation in France

Roxane Sansilvestri [1,*], Mateo Cordier [2] and Thibault Lescuyer [2]

1 Ecologie-Systématique-Evolution Lab (UMR 8079), Université Paris Saclay, 91400 Orsay, France
2 Cultures-Environnements-Arctique-Représentations-Climat Lab (CEARC), Université de Versailles Saint-Quentin-en-Yvelines, 78280 Guyancourt, France; mateo.cordier@uvsq.fr (M.C.); thibault.lescuyer@uvsq.fr (T.L.)
* Correspondence: roxane.silvestri@gmail.com

Abstract: International policies promote renewable forms of energy to mitigate climate change. In Europe, the production of electricity using wood biomass represents one of the most popular energy alternatives. In 2012, France initiated a large-scale strategy to develop wood biomass energy. The biggest wood biomass power-plant project has been developed in the French Mediterranean area and its huge size raises several issues for the short- and long-term sustainability of local forests and associated economic sectors. The French Mediterranean forests provide four types of economic goods (private, club, common, and public goods) and multiple ecosystem services, which makes them complex to manage under an energy transition policy. In this paper, we applied three qualitative methods, namely interviews, participative workshops, and observant participation, and three conceptual models, namely (i) Ostrom's (2010) self-organization key conditions, (ii) the types of economic goods classified according to their excludability and rivalry properties, and (iii) the ecosystem service categorization system of the Millennium Ecosystem Assessment (2005). With our methods, we show that the renewable strategy chosen in France replicates the current centralized production model based on fossil and nuclear fuels. Thus, we demonstrate that European, national, and local authorities fail to consider the multiple ecosystem services that forest management strategies should include to face the energy transition, climate change, and the other ecological challenges of the 21st century.

Keywords: common-pool resource management; local vs. global; economic oligopoly; panacea paradigm; renewable energy; sectoral organization

Citation: Sansilvestri, R.; Cordier, M.; Lescuyer, T. Winners and Losers in Energy Transition: Study Case of Wood Biomass Power-Plants Implementation in France. *Forests* **2021**, *12*, 1139. https://doi.org/10.3390/f12091139

Academic Editors: Guangyu Wang

Received: 24 June 2021
Accepted: 22 August 2021
Published: 24 August 2021

1. Introduction

In the face of climate change and the energy crisis, international policies seek to accelerate renewable and low-carbon energy development (e.g., the "Paris agreement" held in 2015). One such renewable resource is wood biomass, proposed as an alternative source of energy to reduce greenhouse gas emissions and create a new bioeconomy market in the energy sector [1–4]. However, the development of renewable and low-carbon energy raises crucial issues concerning the capacity of societies to implement this energy transition. We defined the energy transition as a political strategy, described by Griffiths (2019) [5] as the aim to change our energy and economic system throughout the remainder of this century by a shift from reliance almost entirely on fossil fuels to a much greater reliance on renewable energy [6,7], especially at a local scale.

The European Directive 2009/28/CE on the promotion of the use of energy from renewable sources (hereinafter named H2020) sets the common EU objective to reach at least 20% of renewable sources in terms of gross final consumption of energy by 2020 and to develop bioeconomy markets that rely on renewable biological resources [8,9]. Accordingly, policymakers are facing the difficult challenge of developing new renewable energy sectors and associated bioeconomy markets under high uncertainty while taking into account the three sustainability pillars: the ecological, the social, and the economic. Yet, the literature shows that many political instances of energy transition planning fail to consider some issues, such as landscape and cultural values, human health, and either equity or justice issues [10], as well as fail to consider environmental issues in developing new bioeconomy sectors [2,11,12].

In France, reaching the H2020 goal requires that renewable energy sources must more than double their 2005 share (10.3%) in terms of gross final consumption of energy to reach a final goal of 23% at the national scale (each member state has a different goal, which allows a 20% share to be achieved at the European scale). In 2006, the wood biomass energy in France represented 83% of the thermal renewable energy consumed by households and industries (i.e., 9 million tons of oil equivalent (toe) of wood biomass per year). This represents 50% of the French renewable energy production [13,14]. Moreover, wood biomass energy is considered as an interesting strategy for France, taking account that the standing volume of wood in France has been increasing by 0.7% per year since 1992 [15]. Even if an increase of the wood standing volume does not necessarily increase its availability, in 2016, the French Forest Ministry edited a national report to promote forestry sectors with a strategy to develop wood biomass energy at large scale [16].

In the National Strategy for Ecological Transition towards Sustainable Development (2017) report, France defines bioeconomy as "the photosynthesis economy, and more generally, the living world economy. It encompasses all biomass production and processing activities, wether forestry, farming or aquaculture, directed at the production of food, feed, bio-based products and renewable energy" ([17], p. 39). In its bioeconomy strategy, France considers wood biomass as the main pillar [8,18]. As many authors highlighted, environmental or ecological issues remain vaguely addressed in bioeconomy sector development [12,19,20]. Moreover, Dietz et al. (2018) said that future research in bioeconomy and new renewable energy markets should contribute to documenting examples of implementation processes and outcomes of new sectors at local scales. The wood biomass energy sector fits in the bioeconomy strategy, and thus it is crucial to analyze the implementation of the process to identify failures and keys to success.

In July 2010, the Energy Regulatory Commission published its 4th Biomass call (hereinafter named CRE4) to develop renewable energy plants in France. In February 2011, 15 projects were accepted, such as the project of the German Company E-On/Uniper, which proposed to convert a coal-fired plant into a wood biomass power-plant. The project is located in the region Provence-Alpes-Côte-d'Azur (PACA), which corresponds to the French Mediterranean area. Despite the fact that several of the CRE4 specifications required in the call were not respected, the project was accepted by the public authorities because of its capacity to contribute to European and national renewable energy targets and to help the French government make progress in closing coal-fired power-plants. Moreover, the project has been supported by regional authorities for its capacity to improve regional energy security and to either create or protect regional jobs. The French government invested €66 million, and the E-On/Uniper company invested €250 million, for the new wood biomass plant installation.

It is important to note the huge size of the E-On/Uniper project, and thereby its massive potential ecologic, social, and economic impacts. The CRE4 call estimated that, in the coming years, 420 MW of additional energy capacity should be created to supply the ever-increasing energy demand in France. Several companies have won the CRE4 call, but the E-On/Uniper project covers 36% of the 420 MW. The other companies that won the CRE4 call have presented projects that will produce a much lower percentage. For example, the second biggest project that won the CRE4 call will have an energy capacity of only 26 MW. Considering the power capacity of the E-On/Uniper plant and the required supply, approximately 850,000 tons of wood per year, this project raises many issues for the short- and long-term sustainability of the PACA region in respect of the local economy [4,21–23], biodiversity and forest ecosystem services [24,25], and the capacity of local actors to organise the regional sector [26,27].

The capacity of actors to organise themselves at the regional scale (the PACA region) to fulfil biomass policies decided at a larger scale (EU) is crucial to diminishing the costs (economic, social, or ecologic) and the potential negative impacts ([27], p. 3). The arrival of the power-plant is putting pressure on wood resources and represents an ecological risk. Development of the forestry sector through reliance on E-On/Uniper, a multinational group with 12,000 employees worldwide (Uniper, 2018), is not conducive to employment. Very large-sized companies (more than 250 employees) are known to create fewer jobs than do small- and medium-sized companies (between 50 and 250 employees) ([28], p. 14).

In this paper, we analyse both emergent trends during the implementation of a new energy sector at the regional scale and the preliminary organization of regional stakeholders to respond to national strategy. Our analysis is based on the study case of the E-On/Uniper power-plant in the PACA region. Our approach considers the local environmental and historical conditions, the current stakeholder's network, the type of goods and services offered by forests and the political context. To grasp the complexity of the social–ecological system, we carried out document analyses, semi-structured interviews, participative workshops, and local immersions in the form of observant participation [29,30].

Thus, we answer three research questions: (A) What are the historical characteristics framing the capacity of local actors to self-organise to manage forests? (B) How are local actors trying to self-organise at the medium scale to respond to the top-down energy policy decided at the large scale? (C) Is the implementation of the energy transition policy by the E-On/Uniper power-plant, improving local forest management and sustainability?

2. Materials and Methods

2.1. Case Study: The French Mediterranean Region and Its Forestry Sector

The study was conducted in the PACA region, delimited by administrative limits (The authors choose to circumscribe the study case to the administrative limits. We are aware that it is questionable, but this decision was made because available data are inventoried at the administrative scale of the PACA region, and many reports are edited at this scale too), which covers 31,400 km^2 and is the most forested region in France. Forests cover 1,606,000 ha, which represents 48% of the regions surface area [15]. The forest cover in the region has been increasing by 1.35% per year since 1985 [15]. However, even though the forest cover is one of the largest in France, the productivity of Mediterranean forests remains low, at 86 m^3/ha against 175 m^3/ha, on average, for the rest of France. The wood productivity is limited by severe climatic conditions, frequent fires, and poverty of soils in some zones [31–33].

From the forestry sector and political stakeholder point of view, the PACA forest are considered as underexploited, because the "harvest forest biomass production" ratio is approximately 20% in PACA, whereas the national average is estimated to be between 56% and 65% [15]. Moreover, the soil characteristics, the slope, and road accessibility are not suitable for forest exploitation, which explains why 70% of the volumes of standing timber (80 million m^3) in PACA are considered difficult to exploit [34].

The power capacity of the E-On/Uniper woody biomass power-plant amounts to 150 MW per year and is aimed at supplying electricity to 440,000 households. The facility is designed as an incinerator (specific French legal status) and is therefore also capable of burning different types of wastes, such as green waste, recycled wood, and some household wastes. The project has received considerable criticism, particularly because of its overall low efficiency of 44% due to the lack of a heat recovery system (although the CRE4 call for proposals required an efficiency of 60% minimum). The amount of resources burnt in the plant will be very large and equivalent to 850,000 tons of biomass per year, including 600,000 tons of wood that are currently mainly imported from abroad (60%) but should come from within 400 km of the plant by 2025. This project adds up to two existing plants in PACA: a paper pulp mill, named Fibre Excellence, using 1,150,000 tons of wood per year and another woody biomass power-plant, named Inova, which has a power output of 22 MW and uses 150,000 tons of forests product per year from a supply radius of 150 km.

2.2. Interview Design and Key Actors' Selection

To develop our research hypothesis, we used an inductive methodology related to the grounded theory. Hence, our approaches are based on poststructuralist and constructionist methods of social sciences. The poststructuralist framework helps us to define the conditions of emergence and the current properties of the PACA social–ecological system. The constructionist framework helps us to determine the individual motivations and perceptions of actors [35]. These two methods allow to understand the social–ecological system (e.g., actor's network) and its current processes (e.g., energy transition process).

To collect information on local conditions and understand all the underlying dynamics of the PACA forest sector, we used four methods (Table 1): (1) document analyses (media, archives, maps, official reports, statistics, etc.), (2) four local immersions in the form of observant participation [29,30] during a total of 18 weeks over the period 2014–2018, (3) semi-structured interviews of 1–2 h each carried out with a total of 40 stakeholders between April 2017 and June 2017, (4) and three participative workshops of 3–4 h each with a total of 12 local stakeholders carried out in April 2018. The interviews and participative workshops have been transcribed and interpreted by the authors in accordance with the traditional interpretive method used in social science research [36], respecting the point of view of the interviewees.

Table 1. A categorisation of information types and methods used to analyse them. Note: tacit information is known only by individuals, whereas explicit information is shared, with some level of agreement. The 3 shaded cells correspond to information types that we did not use for this study. (Adapted from Fabricius et al., 2006 [37]).

	Tacit Information	Explicit Information	Methods Used
Formal information	Private images or photographs	Ecosystem assessments	Document analyses (archives) and local immersions
	Unpublished models and databases	Peer-reviewed papers, chapters, or books in the scientific literature	Document analyses (official reports, medias)
	Diaries	Peer-reviewed databases	Document analyses (models, statistics)
Informal information	Opinions	Oral traditional knowledge	Semi-structured interviews and participative workshops
	Experience	Indigenous knowledge, rules, and practices	Local immersions and semi-structure interviews
	Intuition	Communal beliefs and values	Semi-structured interviews and participative workshops
	Private beliefs and values	Untested scientific databases	Local immersions, semi-structured interviews and participative workshops

We interviewed three types of actors: industries/companies (economic actors), institutions/associations (political actors), and private forest owners (private/citizen actors). We designed distinct sets of questions for each type of actor. Based on our expertise of the forest sector and the PACA region, we chose the first key stakeholders to be interviewed, and additional key stakeholders to be interviewed were either recommended by the first ones or simply mentioned by other stakeholders in informal conversations during observant participation.

Participative workshops have been organized with several stakeholder types as interviews (the same listed in previous paragraphs). Note that E-ON/Uniper and FibreExcellence participated in these workshops, contrarily to INOVA that did not respond to our invitation. The goal of workshops was to identify and design narrative scenarios for the forest and forestry sector of the PACA region, based on experiences, skills, perceptions and expertises of stakeholders. The main question addressed to actors at the beginning of the workshop was "What is the future for the forest/forestry sector in PACA region?" The workshop took place in three successive steps as described below.

Step 1: Identifying main issues for the PACA region. Each actor expresses his main concerns about forest and/or forestry sector, and categorizes them as risks, opportunities, or priorities. All issues are presented to the group. Then, each actor is invited to identify the most critical or worrying issues by sticking a red sticker. At the end of Step 1, we eliminated issues with the lowest number of stickers which allowed us to identify three main issues for the PACA region in each category in order to have: a risk, an opportunity, and a key priority issue.

Step 2: Highlighting components or dynamics which influence these issues. Each actor defines one driver for each issue and specifies what type of influence it is, negative or positive. Then, similarly to the previous step, actors identify the drivers with the strongest influence sticking a red sticker. At the end of Step 2, we have two drivers for each issue, one negative and one positive.

Step 3: Determining the capacity of each actor and the uncertainty. Each actor identifies drivers for which he/she could have a capacity for action, and they propose several action plans. Then, with a scale from 1 (low uncertainty) to 5 (high uncertainty), each actor sticks a number on the previous action capacity to assess the perceived uncertainty.

Step 4 is the last one of the participative workshops and it consists of drafting a scenario for the PACA region according to work carried out by actors in Steps 1–3. We drafted one business-as-usual, one probable, and one desirable scenario.

2.3. Conceptual Framework: Key Conditions for Self-Organisation in Managing Common-Pool Resources Considering the Complexity of Good Categories in Forests

Common-pool resources are goods, manmade or natural, that are large enough that their exclusion from the resource system is costly (and sometimes impossible) and consumption of a resource unit is rivalrous (i.e., there is rivalry with other consumers, the resource is no longer available to others after consumption by someone) [38–40]. These two characteristics make common-pool resources susceptible to overharvesting and destruction. In a property system, common-pool resources are not free to access, contrary to public goods, and property rights could exclude certain categories of actors. However total exclusion is rare in our case. For example, although rules might exclude access to forests for trekking, if there are no fences, which is often the case in huge forest areas, such exclusion is not fully operational. Some common-pool resources such as forests are simultaneously rivalrous and non-rivalrous, depending on the resource unit considered, i.e., the ecosystem service [39]. This shows that some ecosystem services supplied by forests could be classed as belonging to the category of public goods, which might change the way they are managed.

For a good to be properly managed in a way to satisfy supply capacity and demand, it is necessary that the property rights of the good are well-defined. However, the forest is a peculiar good in the sense that it provides multifunctional services and associated resources for which property rights are not well-defined or well-known. In such a case, four types of goods co-exist, i.e., private, public, common, and club goods, without clear legal regulation considering them, which does not ease their management [38,40,41]. The ecosystem services offered by forests could be sorted in the four categories (Table 2). From this statement, we consider forest as a boundary object, which makes management practices complex to implement, and by extension, challenges the development of new bioeconomy sectors [42].

Table 2. The different types and levels of public and private goods related to forest resources and services. These types have important economic and social implications.

	High Excludability	Low Excludability
High rivalry	**Private goods** *(Provisioning ecosystem services)* Resources sold on a market such as wood, meat, fungi, genetic, and biochemical resources.	**Common goods** *(Provisioning and recreational ecosystem services)* Stocks of woods, resins, fruits, games. Recreational activities such as walking, biking, jogging, camping.
Low rivalry	**Club goods** *(Recreational, regulating and provisioning ecosystem services)* Services provided by hunting associations, forest cooperatives, forest associations.	**Public goods** *(Regulating, supporting and recreational ecosystem services)* Cultural services such as landscape esthetical value. Regulating services such as fire, flood, climate, and pollution regulation services. Supporting services such as biodiversity support.

Forest overharvesting or mismanagement, like many other environmental problems conceptualized as "global problems", are the cumulative result of actions taken at diverse levels, i.e., at the local level, by regional and national authorities, and by international institutions [27]. Solving this problem requires collective action, and many actors at diverse levels need to change their behaviour. Ostrom (2010) identified at least six key conditions required for self-organisation to succeed in solving collective action problems (see Box 1). In the following sections, we use the conceptual framework from Box 1 to present and analyze our results. In this paper, we have considered the local scale for cities, villages, unions, associations, private firms, and communities; the regional scale for regional authorities; and the large scale for national and European authorities and institutions.

Box 1. Key conditions required for local stakeholders to self-organise to solve collective action problems in managing common-pool resources [27].

1. Reliable **information** is available about the immediate and long-term costs and benefits of actions.
2. The individuals involved see the **common resource as important** for their own achievements and have a **long-term** time horizon.
3. Gaining a **reputation** for being a trustworthy reciprocator is important to those involved.
4. Individuals **can communicate** with at least some of the others involved.
5. Informal **monitoring and sanctioning** (award and punishments) is feasible and considered appropriate.
6. **Social capital and leadership** exist, related to previous successes in solving joint problems.

3. Results and Discussion

3.1. The Weight of History: Absence of Long Term Vision and Leadership

Despite there being a very large forested surface area, the inhabitants of the PACA region do not have a forestry culture (more details in Sansilvestri et al., 2020) ("There is no culture for good forest species and noble wood in Mediterranean area" (Forest expert from private office, personal interview, May 2017)), which does not help forest actors involved in the wood supply chain to see forest resources as a common-pool resource for their own achievements over a long-term time horizon. In other works, key condition No 2 (Box 1) is not fulfilled. This lack of common view can be explained by several historical and environmental conditions.

As with many Mediterranean climate regions, PACA is exposed to a high fire risk, which creates fear among the community of forest owners and pushes them to focus forest management on reduction of the fire risk (Box 2). Thus, the forest practices considered as good and subsidised by public authorities focus chiefly on fire defense ("Here, when you want to sign a management plan with a private forest owner, you just have to talk about fire. They are so scared that they accept any management plan" (forest expert CRPF PACA, personal interview, April 2017); "I think that forest management should focus on fire risk decrease" (private forest owner, personal interview, April 2017)), discarding any other kind of longer-term objectives (Table 3, "fire-defense" strategy). Actually, the fire risk and fear are exacerbated by the pressure of climate change due to increases in the intensity and duration of droughts in the PACA region [31–33]. The fear of fire and the real risk enhanced by climate change, combined with the low value of wood, are three factors explaining why fire-defense is seen as the most important practice by private forest owners. The fire-defense strategy was presented as a business-as-usual scenario during the participative workshop identifying the impacts on the different forest ecosystem services (Table 3).

Table 3. Description of the different scenarios and the associated impacts on the four ecosystem service categories, and on the accessibility to forest goods and services. The different scenarios have been developed with stakeholders during participative workshop, and then specified by our analysis based on interviews and document analysis. The fire-defense and the conservation strategies have been described as business-as-usual scenarios, the biomass policy (at large and local scale) as the probable scenario, and the fire-management strategy as the desirable scenario by the majority of actors during participative workshops. Legend: the arrows indicate the increasing, decreasing or no change evolution of the characteristics listed in the first column (large arrow indicates a strong increase or decrease). The color of the cells indicates the desirable level according to the assessment of stakeholders during workshops (Dark green cells = extremely desirable for future, light green cells = desirable, yellow cells = no assessment or neutral, light red cells = no desirable, dark orange cells = extremely desirable).

Ecosystem Goods and Services, and Economic Competition Characteristics	Fire Defense Strategy	Conservation Policy	Biomass Policy (Large Scale)	Biomass Strategy (Local Scale)	Fire Management Strategy
Provision goods (for wood only)	↘	→	**↗** [1]	↗ [2]	→ [3]
Supporting services	↘	↗	↘	→	↗
Regulating services	↘	↗	↘	→	↗
Cultural services	→	↗	↘	↘	→ [4]
Rivalry	↗	↗	↗	→	→
Excludability	↗	↗	↗	→	↘
Unfair economic competition	↗	↗	**↗**	→	↘

[1] Biomass policies (at large scale) increases the provision of wood on the short term but it is likely to generate a decrease on the long term; [2] Biomass strategies (at local scale) have to remain limited on the long term to maintain the service of wood provision; [3] The provision of wood could increase on the long term under fire management strategies if public knowledge about forestry and forest experts' education are developed. [4] Cultural services could increase on the long term under fire management strategies if the policy-decision process includes landscapes management at the scale of the territory and all of the actors (foresters, shepherds, inhabitants, associations etc).

Box 2. Details about fire defense vs. fire management strategy.

Fire-defense: The increase of human pressure in the Mediterranean basin across century drives policy-process to develop strong fire-policy to protect human populations. But, Mediterranean forests have natural cycle which include fires, e.g., some species need fire for reproduction or expansion. Today's management practices focus on firefighting, to avoid or to constraint fires. Such practices are named hereinafter *"fire-defense management"*. The traditional practices in *fire-defense management* aim to create corridors in forest by cutting trees, hard undergrowth clearing and permanent fire monitoring units.
Fire-management: Also called "prescribed burning", it consists in voluntarily burning some forest areas to improve socio-ecological system management. For now, prescribed burning is underused in the Mediterranean basin, it is applied on only 3% of forest lands [43]. Fire-management practices plan the use of fires in the SES to achieve clear objectives such as biodiversity preservation, fire risk reduction or traditional landscape protection [44,45].

Though actors at the regional scale all agree on the fire-defense service as being the forest public good to be managed, this is not the case at upper levels. Considering that the Mediterranean basin is one of the biggest biodiversity hot-spots in the world, most regulations decided by decision-makers at European and national scales have been implemented on a regional scale to protect biodiversity. Yet, the development of biodiversity conservation regulations in the PACA region since the 1990s has preserved its high natural capital (natural capital is defined as a stock that yields a flow of services over time [46] and as the biological components and intrinsic functioning capacities of the ecosystem [47,48]), constrained its economic and social capital (social capital is defined as the social networks, cognitive elements, values and perceptions, and culture of the human system [49–51]) that rely on the forest ecosystem). Among the various biodiversity regulations in PACA, there is the Natura 2000 ecological network (Birds (79/409/CE) and Habitat (92/43/CEE) EU Directives), a man and biosphere reserve, and eight regional protected parks (Figure 1). Each protected zone strongly limits human activities, forest operations included. Hence, a large part of the PACA surface is protected. Preserving forest biodiversity allows generation of many ecosystem services, but they are hardly economically valuable for foresters ("From societal point of view, forests have to be conserved. No possibility to talk about harvest operations or money here" (Forest cooperative member, April 2017)). Therefore, under such regulatory pressure for biodiversity set at upper scales, local foresters could never develop their activities properly in a well-structured forestry sector. This has generated a lack of forest practice knowledge and has impeded the emergence of successful regional forestry activities (see the "Conservation policy" strategy in the Table 3, which provides a high level of supporting services but a low level of provision services). The strategies "Fire-defense" and "Conservation policy" represent the historical trends of the PACA region and can be defined both as business-as-usual scenarios for the region.

It was only in the 1950s that a big company, named Fibre Excellence, settled in the PACA region to produce paper pulp (Figure 1). Fibre Excellence is still in operation. Indeed, this company generated a boom in the forestry sector that coincided with the large-scale abandonment of agricultural land and forest area expansion. However, economic profits in the forestry sector continued and remain underdeveloped because of the lack of diversity in wood processes and markets (Figure 2). The current lack of management plans and secondary processing sectors keep the wood price per ton in the PACA region at a very low level, forcing forest owners and foresters to produce wood in terms of quantity, rather than quality ("Here, and in France in general, we know how to produce wood, but we don't have wood processing industries. This does not encourage to develop management plans for future" (Forest community association member, personal interview, May 2017)).

Until 2011, Fibre Excellence developed the industrial forestry sector of PACA based mainly on production of paper pulp (a low added value product). However, recent national energy transition goals involving the implementation of two wood biomass power-plants in the PACA region, E-On/Uniper and Inova, changed the deal. The Inova company built a new wood biomass electric power-plant in 2014 and opened it in 2016. Meanwhile, the E-On/Uniper company proceeded to convert an old coal-fired power-plant in 2014, which did not open yet (Figure 3). Fibre Excellence is continuing its activity in the PACA region, but it now has to "play" with two new economic competitors for wood resources ("The INOVA project could have been a good opportunity for our region, but with E-ON/Uniper and INOVA, it will be complicated" (Forest expert from CRPF, personal interview, April 2017)). Today, these three companies create an oligopolistic wood market in the PACA region.

Along with power-plants implementation by national authorities, several local actors have tried for many years to develop consistency in the forestry sector through the development of a collective heating network with community boilers at local scale (described as the "Biomass strategy (local scale)" in Table 3). This can be observed through the growing number of regional small community wood-fired boilers, which increased from approximatively 30 in 2003 to 284 in 2016 [34]. However, the lack of leadership in forest sector and the unstructured network limit the organization of actors as reflected by actor's assertions during interviews ("Communication is very difficult, there are too many interlocutors" (Forest operator, personal interview, May 2017); "The interprofessional structure is too young to create cohesion between actors" (Foresters union member, personal interview, April 2017)). All the historical traits mentioned above result in low social capital and no common vision of the regional resource and long-term achievements. Consequently, Ostrom's key conditions No 2 and 6 are not fulfilled (Box 1).

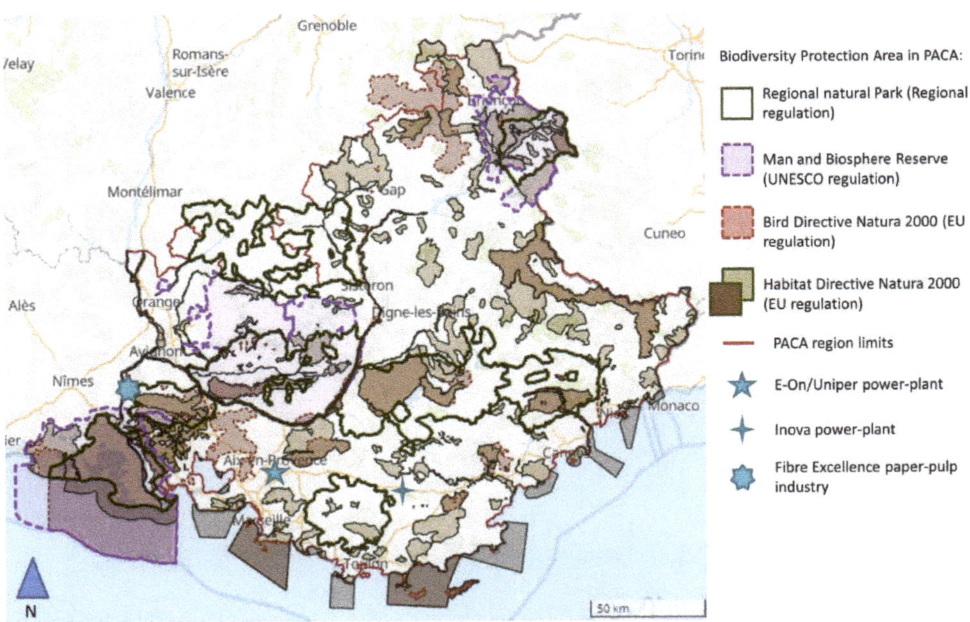

Figure 1. Map of PACA environmental regulations decided at upper scales (global scale: man and biosphere reserve labelled by UNESCO; European scale: Natura 2000 area; National and regional scales: Regional Natural Parks) and location of the three main industrial actors (E-On/Uniper, Inova and Fibre Excellence). Map edited from the online Regional Natural Park database for geographical information system (SIT PACA Database).

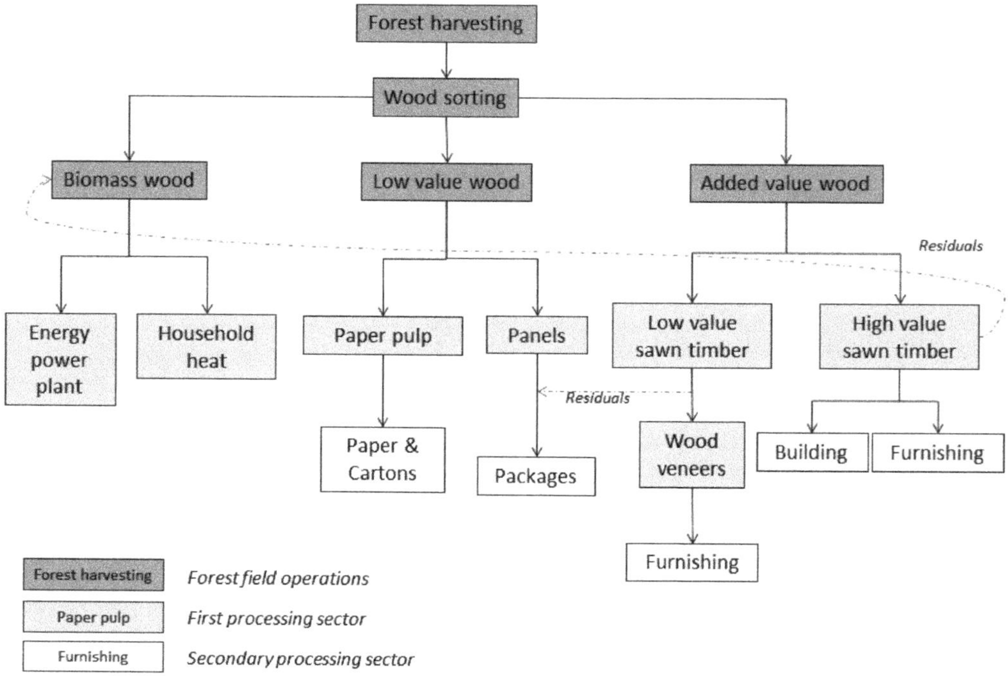

Figure 2. Simplified description of a structured and diversified forestry sector. As it was stated by most of PACA forest actors "*A more profitable and structured forestry sector should involve a regional wood supply chain with both first and secondary processing sector of transformation*". Yet, the development of a secondary processing sector requires wood with higher added value, necessitating management plans for selection of young trees with high potential value, wood sorting once trees are cut to select high value lumber, and technical structures.

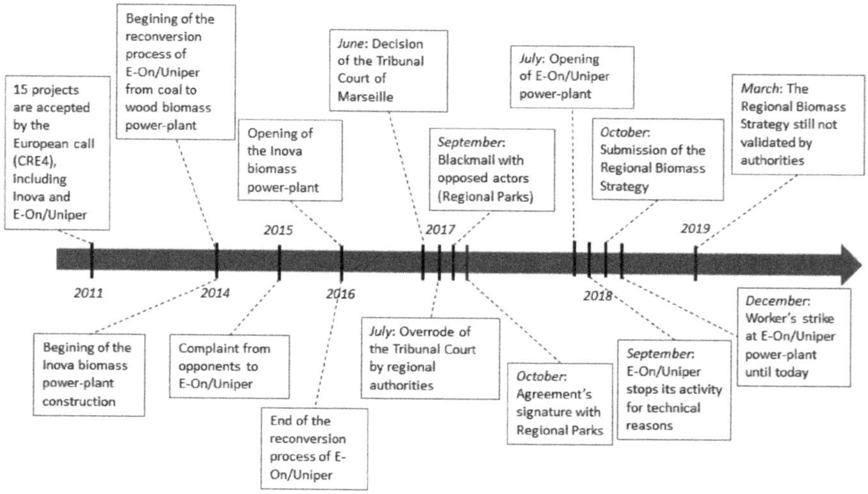

Figure 3. Chronology of the arrival of woody biomass mega power plants in the PACA region.

3.2. The Local Response to Top-Down Policy: Consequences of Lack of Information and Trust

The E-On/Uniper and Inova power-plants offer economic and political opportunities for the region. However, at the same time, the top-down implementation of the national woody biomass policy has sparked considerable disruption in the regional governance and for local foresters and citizens ("It is a big project suddenly launched in a very fragile context" (Forest community association member, May 2017)).

As explained in the previous section, regional forest actors have developed a collective heating network of small boilers with short wood supply chains. Hence, the implementation of the large E-On/Uniper power-plant in the region has been perceived as treason by foresters and regional decision-makers who work hard for the forest sector. Local actors perceive the wood energy approach as being probably more in line with the decentralised energy system imagined by Rifkin (2011) [52], an approach deeply rooted in the local territories breaking with the conventional centralised energy production based on huge energy power-plants. The giant E-On/Uniper company requires a much higher amount of wood resources, and has far greater harvesting capacities, than do local foresters. This situation drove several actors to prosecute E-On/Uniper in 2015 because of a weak environmental report with no consideration for regional actors, the context, and the forestry network that was already in place.

The political stakes of the E-On/Uniper project skewed the actor map and contributed to imbalances in the power involved. In June 2017, the Administrative Court of Marseille announced its decision to cancel the authorisation to operate that the French State had granted to E-On/Uniper in 2012. This decision was based on the lack of sufficient environmental impact studies, mainly on overharvesting risks for the forest, for the economic sectors present in the region before E-On/Uniper installation, and on transport pollution and noise disturbances. Thus, the Administrative Court of Marseille required E-On/Uniper to produce new environmental impact study that took into account its 400 km harvest's operation radius and not only the 3-km legal radius. However, a few weeks after the decision of the Administrative Court of Marseille, the political representative of the national government at the regional scale (named Préfet in French) overrode the decision and gave to the E-On/Uniper power-plant temporary authorisation to operate for nine months. Finally, in September 2017, two regional natural parks (the park of Luberon and the park of Verdon), who were the main actors to sue E-On/Uniper in court, were "blackmailed" (wording used in the press article) by the Préfet [53]. They were forced to sign an agreement with E-On/Uniper, otherwise they would not receive public subsidies from the regional authorities, the main source of their income. This last action highlighted the key motivations of this industrial project, motivations that are essentially political, questioning the real considerations for the interests of the PACA region, its forests, and its forestry sector. As demonstrated by Upreti and Van Der Horst (2004) [54], the non-negotiated and inflexible top-down implementation of the strategy, namely imposing social, economic, and environmental costs not previously discussed locally, has been interpreted by local people as a case of national targets having supplanted local issues ("I don't think that all of this project has been thought inside the PACA region. The reactions started at the arrival of E-ON/Uniper" (Forest operator, personal interview, Avril 2017)). This hard top-down implementation blocks the development of a trustworthy relation with a bad reputation for E-On/Uniper, implying that key condition No 3 is not fulfilled (Box 1).

Due to this preliminary reluctance and the weak network organisation of forest actors, the delays observed in operational planning gave decision-making power to big industries and created insecurity for small local forest companies. To respond to the industrial demands for wood resources, the PACA region had to create a committee, named the Regional Biomass Committee, in charge of implementing a supply strategy. The committee is constituted from regional institution members, a private forest owner union, a local forester company union, the forest cooperative, and all wood industrial actors, including E-On/Uniper, Inova, and Fibre Excellence. However, the committee was established after the national decision allowing E-On/Uniper and Inova to set up in the region, and the final strategy report, named Regional Biomass Strategy (RBS), was published only in October 2018, whereas the development of the wood biomass sector for electric energy purposes began in PACA in 2014 (see the chronology in Figure 3). In the short-term, the delay in the RBS has created economic insecurity for forest operators and reluctance from private owners. In the long-term, it could threaten forestry sector development in PACA, the local biodiversity, and the carbon storage potential, so by extension the sustainability of PACA's forests. Such a lack of reliable information illustrates that the key condition No 1 is not fulfilled (Box 1).

The first impact of the RBS delay is an economic one with high disturbances in the forestry sector ("We created competition between wood biomass and the other wood sectors" (Forest operator, personal interview, June 2017); "The arrival of E-ON/Uniper increased wood prices very fast and created a bargain effect for foresters. Today, it is another story ... " Forestry union member, participative workshop April 2017)). Even if the wood biomass development had a deadweight effect for regional economic actors, today profits for foresters remain modest. This can be explained by several factors. First, in the absence of a coordinated strategy, most regional forest entrepreneurs have invested independently in hard mechanization (We define "hard mechanization" as large and heavy equipment such as felling machines, chippers, or skidders. All of these machines are very expensive and bulky. This equipment is recent, until now most of foresters used light harvesters or chainsaws, and trailers), anticipating the future wood demands ("Today, each forest operator, even the smallest one, has its own mechanic harvester, skidder or truck" (Forest operator, personal interview, June 2017)). One mechanical harvester costs approximately €600,000, which explains why many forest entrepreneurs are now indebted. Second, E-On/Uniper has a constant need for biomass. At the beginning of the contract, E-On/Uniper establishes prices and biomass supplies with forest entrepreneurs for five years, and they vary according to E-On/Uniper's annual financial capacity (approximately €65 million per year, E-On/Uniper assertion, June 2017), the energy global market, and economic arrangements between E-On/Uniper and the French Electricity Company EDF (The French Electricity Company (EDF) is the first producer and supplier of electricity in France and Europe). Fibre Excellence does not establish five-year contracts. It buys wood logs to forest entrepreneurs when it needs it following the current market price and not the fixed price of the contract. Thus, wood biomass resale prices to the power-plant do not increase, even if the standing wood prices increase (see "Biomass-Policy (large scale)" in Table 3, which generates high provision services but also high rivalry). This situation obliges forest entrepreneurs to harvest large quantities to generate adequate salary and profits (e.g., to reimburse their debts) while the traditional wood market (e.g., sawmills cutting wood for construction and equipment purposes) would valorize wood quality ("Forest entrepreneurs seeking to maintain the stability of their activity will prioritize the supply contracts they signed with wood biomass power-plants. This is ruining the efforts developed to value woods for construction and equipment purposes, while the sawmill sector is in so much need of revival" (Richard Fay, regional forest association, Pers. Comm., 2017); "They take our valuable wood to burn it. There is no competition, we are just disappearing" (Sawmill owner, personal interview, May 2017)). Moreover, this need to produce wood in quantity threatens the local biodiversity and PACA's landscapes.

A second consequence of the RBS delay is the mistrust felt by private forest owners towards the forestry sector, and especially the wood biomass energy one. In fact, 65% of forests in PACA are private properties that are fragmented into small plots, with 217,850 owners [34]. Moreover, only 20% of the private forest plots have an official forest management plan. In France, only forest plots larger than 25 ha require an official forest management plan certified by the regional institution representing the National Forest Ministry. To mobilise biomass quantities, forest operators must deal with the fragmentation of private forest lands. Hence, the most accessible and smallest plots suffer under pressure from forest operators because operation costs are low on easily accessible plots, and environmental regulations are weak on plots smaller than 25 ha. Thus, we are already observing a dichotomy in the PACA forest landscapes, with some plots being overharvested and others left unharmed. Without a forest management plan, there is no selection of standing trees. Thus, high value trees are not preserved to keep them growing and sold them later to second transformation industries in the future (Figure 4). Given the high speed of development of the wood biomass energy sector, private forest owners feel that their forest resources are "looted" (information collected during participative workshops), which creates resistance from forest private owners and reduces trust, making it difficult to achieve the key condition No 3 in Box 1.

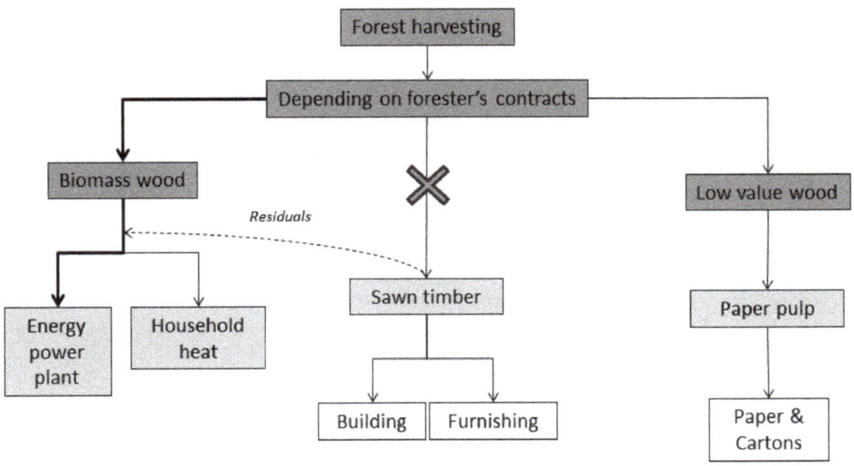

Figure 4. Description of an altered forestry sector with a domination for the biomass wood production. In this case, woods are not still sorting according to their value but according to the forester's contracts with industries.

3.3. Recent Social Capital and Leadership Development: Helping to Solve Joint Problems in Managing the Forest Common-Pool Resource?

Even though local actors questioned the legitimacy of the arrival of the E-On/Uniper power-plant, most of them admitted that a new wood market could create regional wealth and opportunities for the forestry sector ("Now, E-ON/Uniper is here, we have to deal with it" (Forest expert from CRPF, April 2017)). Some local actors understood that this political project was unavoidable and that it could be an opportunity for the development of a regional secondary processing forest sector. In this way, E-On/Uniper and Inova have been a trigger for the PACA region, generating hope for new potential opportunities.

It is a matter of fact that E-On/Uniper has emerged as a potential driver for the regional forest development. The absence of a forestry leader in the PACA region for many years presented an opportunity for E-On/Uniper to endorse the vacant role of leader for the regional forest sector to facilitate the integration of the project inside the territory, which seems to fulfil the key condition No 6 (Box 1). First, in response to the fear of overharvesting, E-On/Uniper engaged in creating a sound forest chart that aims to ensure sustainable forest practices (personal comments collected during participative workshop). Second, for industrial transparency, E-On/Uniper participated in all regional forest meetings and built a close relationship with regional actors. Then, the company went further by planning the installation of a sawmill and wood storage next to its wood-fired power plant to value high quality trees ("We are open to the installation of a sawmill in our structure. Moreover, it will be possible to use "the overflow heat" to dry logs" (E-On/Uniper member, participative workshop, April 2018)). This quelled the controversy raised by several actors (environmental associations, some Forest Entrepreneur Union members, and some Forest Private Owner Union members), who feared that these trees would end up being burnt for energy purposes whereas their height, their shape, and their diameter would have allowed them to be sold to high value-added sectors (Figure 4).

The previous paragraph suggests that wood-biomass power-plants in PACA might have positive impacts on forestry sector for structuration and economic development. However, during the participative workshops, some stakeholders asserted that the E-On/Uniper sawmill will not attract the secondary processing sector in the region: "Today, it is not a problem of resource quantity but of quality. If private owners do not develop a forest management plan in their forest, we cannot increase the value of PACA forests" (Forest Private Owner Union and Forest Entrepreneur Union, participative workshop, April 2018). According to those stakeholders, because of lack of forest management plans on private lands, high value woods are rare, and E-On/Uniper will only be able to value a few high-quality logs that arrive at the plant by coincidence from time to time.

The resulting structuration that UNIPER proposes with its sawmill will be based only on the E-On/Uniper perspective and activity, which is based on biomass exclusively rather than on holistic planning. This monopolistic market could be dangerous for the PACA forestry sector in terms of impacts on prices, resource's supply, salaries, forestry sector development, and employments. In addition, it could be dangerous for the sustainability of local forests because only one stakeholder vision of the forest ecosystem and services is taken into consideration: forest biomass for energy purpose, so the key condition No2 is not encountered. Such a biomass policy at large scale is likely to generate low supporting, regulating and cultural services, as it is described in the scenario in Table 3 developed by local stakeholders during participative workshops. All stakeholders, except E-ON/Uniper, have attested that this scenario is dangerous for the PACA forest if harvest operations continued as they are currently.

3.4. Competition Generated: Loss of Trust and Lack of Monitoring

According to Jenkins et al. (2018) [55], environmental challenges, such as energy transition, could represent an opportunity for the "bottom of the social pyramid" because they involve new stakeholders to create new offers and satisfy currently unmet needs. However, this suggests that all actors, old and new, share at least one common target in the transition process. The European Union sets the H2020 environmental targets and the French government sets renewable energy policies to meet the targets. In terms of climate change mitigation and renewable energy development at both European and national scales, the meaning of this strategy is quite understandable. However, it is questionable if among all the goods and services provided by the forest common-pool resource, we should only consider the public good of climate regulation service or also the other ecosystem services and goods provided by forests at both medium and small scales (Table 2) ("Today, our forest is more vulnerable to wood biomass sector than to climate change" (Private forest owner, personal interview, May 2017)).

To create opportunity for "the bottom of the social pyramid", all actors should be left with the same chances. However, in the PACA region, we observe disparities between large- and small-size forest companies. Large-size companies have much higher technical and financial capacities, in addition to both greater and easier access to forest plots than do small-size companies. As a result, large-size forest companies succeed in selling huge amounts of standing wood to power-plants at lower prices, whereas small local companies do not succeed in harvesting such amounts of wood and cannot afford to sell at low prices. In particular, two companies benefit from the implementation of Inova and E-On/Uniper power-plants in PACA: the Forest National Office (Office National des Forêts (ONF) in French) (The National Forestry Office is a French public institution of an industrial and commercial nature responsible for the management of public forests under the supervision of the Ministry of Agriculture, Food, and Forestry and the Ministry of Ecology, Sustainable Development, and Energy) and the Forest Cooperative (Coopérative Provence Forêt in French) (The forest cooperative was founded in 1997 by 18 forest owners of PACA with the aim of optimizing forest management and making better use of timber cuts by sharing their costs and the skills of their forest technicians. In 2016, the forest cooperative counted 2842 members, representing 120,698 ha in the PACA region).

The ONF has a huge advantage over other forest companies; it manages most of the public forests, which are the property of the State or cities (the ONF plays a key role in organizing the "Biomass policy (large scale)" under a situation of high excludability with harmful consequences on several ecosystem service categories; see Table 3) ("The ONF has a stranglehold on the offer and plays with the availability of wood volumes, and they increase prices when we need it" (Forest operator, personal interview, June 2017)). This means that the ONF has prior access to a vast number of forest plots not only in the region, but also all over the country. The competition is as follows: the ONF keeps the most valuable wood, in terms of high-value woods or low harvest costs, to sell it directly to the processing sector (Figure 2), and it sells the lower value standing wood to the local forestry companies, which will have to cut the wood and then sell it to first processing sector. We observe in this context an extreme and unfair economic competition between actors (Table 3; bold arrow in the cell on the 4th column and 8th row). This situation creates many conflicts within the region and increases tensions between local forest entrepreneurs, drastically reducing the level of trust and impeding achievement of the key condition No 3 (Box 1).

The second actor who benefits from the energy transition in PACA is the Forest Cooperative named as "Provence Forêt". It has economic resources, such as financial means, already amortized machinery and equipment, and human capacities. Their human capacities (12 employees and 10 privileged operator partnerships) allow them to prospect forest private owner, which is time consuming. To do so, they use their members' network and their status which gives them trustworthiness. Moreover, these prospections allow mutualisation of harvest operations with several private forest owner plots that are not members of the cooperative, thus decreasing the cost of harvest operations ("The big ones, as Provence Forêt or Fibre Excellence, they have the possibility to hire people who are only dedicated to prospecting. But us, we are already struggling to pay our lumberjacks" (Forest entrepreneur, personal interview, May 2017)). This gives a crucial advantage to Provence Forêt rather than to small forest entrepreneurs usually working with just two or three employees, and hence lacking human resources to prospect private forest owners. The larger mechanical capacities of Provence Forêt make it a good operator for ONF's complex forest plots. Smaller forest entrepreneurs cannot compete.

As we have said above, the ONF company benefits from priority access to all national public forests. Hence, the ONF company could supply the E-On/Uniper power-plant with wood biomass from another French region; especially from North-East France [56] where tree productivity and plot accessibility are better. Unfortunately for small PACA forest entrepreneurs, they have to deal with local wood resources and hard environmental conditions. This generates great levels of jealousy, which decreases solidarity and increases local entrepreneur isolation in PACA ("The ONF does not play fair. They sell plots at expensive prices which are difficult to access and to harvest, which generates high harvest costs. Besides, they keep the most accessible plots for their own harvest activity, which allows them to sell wood to industries with a larger profit margin" (Forest union member, personal interview, April 2017)). Such a situation is destroying the social capital among local forest entrepreneurs (key condition No 6 in Box 1) and the trust (key condition No 3 in Box 1) between small local entrepreneurs on the one side, and large-size companies, as well as the French government (ONF included), on the other side. Today, only 40% of the wood supply purchased by the E-On/Uniper power-plant comes from the PACA region, while the rest comes from other French regions, Spain, and Brazil (assertion during interview with E-On/Uniper).

This situation has strongly degraded the trust that had prevailed for more than half a century between foresters and industrial sectors (mainly with Fibre Excellence). This explains why the key condition No 3 is not encountered (Box 1). Based on the knowledge we gathered through interviews, participative workshops, observant participation, and archive analysis, we designed an integrative causal-loop diagram describing local dynamics of the PACA forest system (Figure 5). The three feedback loops maintain the system in a deadlock process combining the historical barriers (R1 in blue in Figure 5), the top-down implementation of the wood biomass policy (R2 in orange in Figure 5), and the consequences of the created unfair competition (R3 in red in Figure 5).

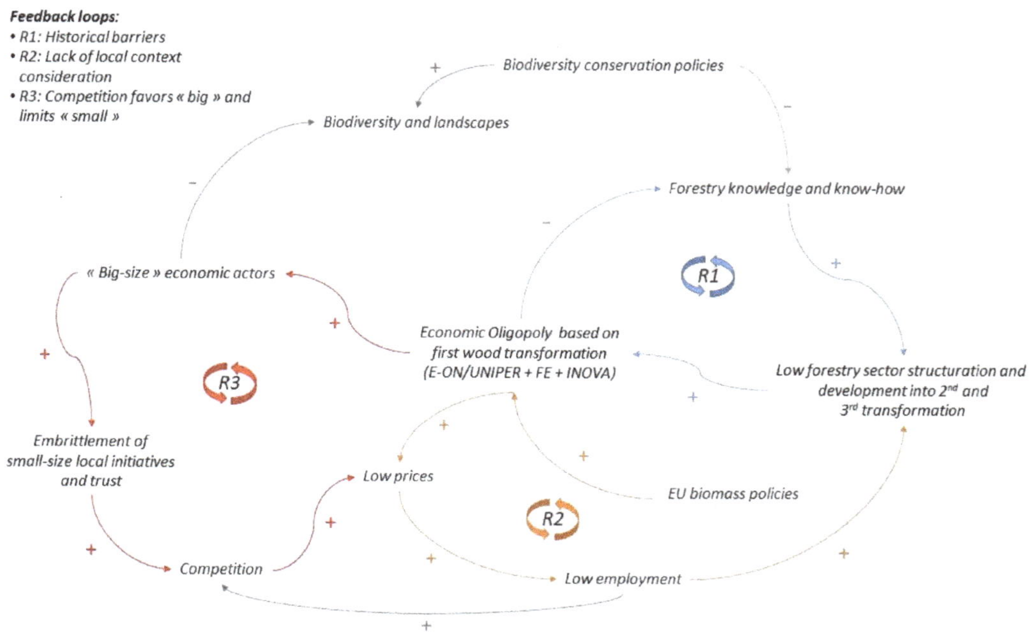

Figure 5. Causal-loop diagram summarizing the local dynamics of the PACA forest system. The sign + describes an enhancing effect and the sign − describes a limiting effect. The blue arrows are associated to the feedback process R1, the orange arrows are associated to the feedback process of R2 and the red arrows are associated to the feedback process R3. The grey arrows represent simple cause-effect dynamics in the system.

4. Policy Implications and Concluding Remarks

We demonstrated that the implementation of the wood biomass energy policy remains difficult considering that the six key conditions (Box 1) designed by Ostrom are not encountered in the PACA's forest social–ecological system. As stated by Jenkins et al. (2018) [55], transition processes without good governance and social considerations create strong inequities, favoring few winners and disadvantaging many losers, as did the historical fossil-fuel transition. The choice of the French government to impose E-On/Uniper and Inova structures in the PACA region clearly generates exacerbated unfair competition, rather than collaboration. This unfair economic competition is generated by two market failures resulting from the centralized wood energy production system chosen in PACA. First, prices of standing wood biomass purchased by E-On/Uniper to ONF and small local forest companies are maintained at low levels due to the oligopolistic market. E-On/Uniper can impose low purchasing prices since there are not many other purchasers in the region under such a centralized energy production system. Second, ONF abuses of its dominant position and does not contribute to market equilibrium, as would occur in pure and perfect competition, since it is unable to reduce its prices of standing wood biomass sold to E-On/Uniper, whereas the 2800 small local forest companies cannot. Thus, E-On/Uniper, Inova, the Provence Forêt cooperative (2842 forest owner members) and ONF are the "big" winners of this energy transition, whereas Fibre Excellence, 226,000 private forest owners, and 2800 small local forest companies suffer from the global wood market and international political agendas.

The transition process needs leadership to create movement inside the social–ecological system [57] and the emerging bioeconomy requires innovative policies to regulate markets [1]. In such a new context, forest actors would be more likely to behave collaboratively to manage their forest goods and services. Here, we propose an energy transition policy based on three steps which would help to solve collective action problems in PACA's forest management. The three steps are developed below and can be summarized as follows: (i) the development of communication and transparency, (ii) the clarification of shared values and common vision, and (iii) the increase of involvement and solidarity. These three steps lead to the emergence of social capital and local leadership (key condition No 6), crucial factors to the success of local energy transition initiatives.

First, forest properties should be visible to all forest stakeholders to create transparency and connection between actors. Transparency would enable the design of an information and communication system in which forest owners would interact and receive information about forest management practices related to their own property as well as information regarding the other forest community members, such as their location, the management plan and practices they choose, etc. (key conditions No 1 and No 4, Table 4). Such a transparent information system open to the members of the forest community would allow comparisons to be made between neighbours. Comparison is important in a collaborative system. It helps community members to learn how to improve their forest management behavior and to develop consideration for other forest owners who have designed a well forest management plan. The learning process operates as follows. Owners without management plan could be informally sanctioned by the community and encouraged to implement a management plan with the help of other members (key condition No 5, Table 4). Owners with proper management plan would receive the expression of informal moral rewards from other members (simply being seen by others as a good person, and the internal satisfaction of doing good). The moral reward is obtained by respecting the norms and the values of the social group that people need to follow to be considered as being a good person who deserves to be liked, respected, or esteemed [58]. Such a collaborative system relies on observations incorporating aspects of behavioral psychology and experimental economics suggesting that the desire to be seen favorably by peers and the belong to a social group is deeply rooted in human psychology [58].

Table 4. Proposition of policy measures for the emergence of Ostrom's key conditions required for local stakeholders to self-organise to solve collective action problems in managing common-pool resources.

Ostrom's Key Conditions	Policy Measures
1. Reliable information on impacts of the collective action	• Modify the regulation to make the cadaster * accessible that would provide concise information on how foresters manage their forests (required by actors during participative workshops). • Make information accessible to forestry institutions concerning the private community, in order to develop collective projects and to design precise cartography (required by actors during participative workshops).
2. A same common resource is seen as important by everyone	• Switch from the concept of common resource (e.g., wood supply) to the common service (e.g., fire management) (identified during participative workshop). Also proposed as "Build upon the entire spectrum of ecosystem services" in the National Strategy for Ecological Transition towards Sustainable Development (2017) [17]. • Identify in a participatory workshop, ecosystem services or goods provided by forest for which all actors commonly agree [59–61].

Table 4. *Cont.*

Ostrom's Key Conditions	Policy Measures
3. It is important to be a trustworthy reciprocator	• This condition is likely to emerge spontaneously once policy measures succeed to build a social group (key condition 2), aware of the costs and benefits of collective actions (key condition 1), communicating and working together (key conditions 4), and sharing the same values (key condition 5).
4. Each one can communicate with the others	• Modify the regulation to make the cadaster * accessible to an association that would be allowed to use the cadastre to create an online platform (proposed by actors during workshops).
5. Informal monitoring and sanctioning	• The information sent to forest owners (key condition 1) will provide a signal on the norms and values shared by the members of the forest community. • Develop labels and certifications of sustainable forest management, as REDD+ [1,11], for all forest plots, not only protected one, to avoid drifts as it is already observed in the region.
6. Social capital and leadership	• Design an environmental tax or a system of payment for forest ecosystem services and transfer the amount collected as an incentive to forest owners who design forest management plans or tax exemption to encourage reinvestments [62]. • This condition is likely to emerge spontaneously once policy measures have succeeded to build a social group (key condition 2), that communicate and learn how to work together (key conditions 1 and 4), and share the same values (key condition 5).

* The cadastre of forest properties is secret in France. Nobody has access except ministry of Agriculture and Forest, regional representative of the ministry, and the Prefet (State's representative in the region).

Second, participative workshops with local actors should be organized to identify shared values and vision of the territory with the aim to favor the emergence of the key condition No2 (Table 4) [59–61], i.e., developing more sustainable management based on a common vision of the social–ecological system (it corresponds to the policymaking key point named "Become sustainable in all dimensions" in the European Commission report on Bioeconomy (2018) [1]. For the purpose of our case study, during the workshops we organized, we identified that many actors (regional parks, residents, hunters, foresters, shepherds, etc.) wanted to preserve the numerous ecosystem goods and services provided by forests (wood logs for multiple purposes; fresh and clean water supply; landscapes; recreational activities; biodiversity; climate regulation; etc.). Accordingly, the fire-management strategy (last column in Table 3) proposed by actors during workshops could be the common vision of the PACA's forest management, considering that this practice could improve also cultural landscapes, biodiversity, social cohesion, traditional knowledge, agro-forestry, pastoralism and security (it corresponds to the policymaking key points "Build upon the spectrum of ecosystem services" and "Enhance cross-sectoral cooperation" in the European Commission report on Bioeconomy(2018).

Third, because some forest goods and services are non-excludable (e.g., common-pool resources in Table 2) and even non-rivalrous (e.g., public goods in Table 2), they benefit to everyone even to community members who do not contribute to forest management costs. Therefore, public authorities should design specific taxes or a system for ecosystem services payment (PES) borne by those using public goods and common-pool resources provided by forests. These taxes or PES would be paid to forest owners in order to encourage them to implement forest management plan respecting shared values and practices (key condition No 3 and No 6) (Statements from participative workshops). In the same way, Elyakime and Cabanettes (2009) [62] propose to decrease taxes for bioeconomy sectors and the creation of investment funds to encourage stakeholders to invest in these mutual funds to reinvest for large scale and collective sylvicultural projects.

The wood biomass energy strategy implemented in the PACA region is based on the panacea paradigm, which assumes that the installation of one sole huge wood biomass power-plant would be able to mitigate climate change (international issue), solve energy demands (European and national issue), and develop the economic forestry sector (local issue). However, many authors highlight that environmental issues and sustainable development require a portfolio of diverse solutions and means, in addition to the creation of new partnerships between companies, civil organizations, and public actors [1,63]. As described by Kleinschmidt et al. (2017) [12] and Pülzl et al. (2017) [18], new bioeconomy sectors such as wood biomass remain mainly a combination of technological progress, markets, and growth, and often forget environmental and social issues. Here, we observe a similar phenomenon, with a strategy that rests mostly on one main, subsidized, and multinational company (E-On/Uniper) and on one forest ecosystem service (biomass provision). This strategy tightens the spectrum of forest ecosystem services and goods which jeopardizes the sustainability of the social–ecological system. More specifically, the creation of such an oligopolistic wood market is risky in terms of the impact on prices, resource supplies, salaries, and employment. Thus, we can observe that France based its energy transition on high tech sector development and emerging industry stimulations with a new power-plant, E-ON/Uniper, as part of the current centralized nuclear energy network [52]. However, a lack of diversity in the social–ecological system reduces resilience in both the mid- and the long-term and makes the system more vulnerable to all types of disturbances.

The case of the PACA region energy transition represents a relevant example concerning the dangerous evolution of forest resource uses during the 21st century at the global scale, namely the hard industrialization of forest management and energy demands. Such a situation raises an important question. What type of forests do we want in the future, i.e., which forest ecosystem services or goods would we like to either preserve or develop?

Author Contributions: Conceptualization, R.S.; data curation, R.S., M.C. and T.L.; funding acquisition, R.S.; investigation, R.S., M.C. and T.L.; methodology, R.S. and M.C.; supervision, R.S.; writing—original draft, R.S.; writing—review & editing, R.S. and M.C. All authors have read and agreed to the published version of the manuscript.

Funding: This research was funded by the Idex Paris-Saclay (Strategic Research Initiatives–IRS) and the BASC (Biodiversity, Agrosystems, Society and Climate) "Partnerships" fellowships.

Institutional Review Board Statement: Not applicable.

Informed Consent Statement: Informed consent was obtained from all subjects involved in the study.

Data Availability Statement: Data are private speeches or narrative discourses, subject to the GDPR regulations. They cannot be published publicly. For more details about data, please contact Roxane Sansilvestri (roxane.silvestri@gmail.com).

Acknowledgments: We are grateful for funding from the Idex Paris-Saclay, which provided support for the postdoctoral position of R. Sansilvestri through the ICSEN project. We would also like to thank Juan Fernandez for his helpful advice on ecological and forestry issues. Finally, we would sincerely like to thank all the stakeholders who participated in the interviews and participative workshops, and gave their time to this project.

Conflicts of Interest: The authors declare no conflict of interest.

References

1. European Commission. *A Sustainable Bioeconomy for Europe: Strengthening the Connection between Economy, Society and the Environment*; European Commission: Brussels, Belgium, 2018.
2. European Forest Institute. *Towards a Sustainable European Forest-Based Bioeconomy—Assessment and the Way Forward. What Can Science Tell Us?* Hetemaki, L., Winkel, G., Leskinen, P., Colling, R., Eds.; European Forest Institute: Joensuu, Finland, 2017; ISBN 9789525980417.
3. Agbor, E.; Oyedun, A.O.; Zhang, X.; Kumar, A. Integrated techno-economic and environmental assessments of sixty scenarios for co-firing biomass with coal and natural gas. *Appl. Energy* **2016**, *169*, 433–449. [CrossRef]
4. Moiseyev, A.; Solberg, B.; Kallio, A.M.I. The impact of subsidies and carbon pricing on the wood biomass use for energy in the EU. *Energy* **2014**, *76*, 161–167. [CrossRef]
5. Griffiths, S. Energy diplomacy in a time of energy transition. *Energy Strategy Rev.* **2019**, *26*, 100386. [CrossRef]
6. Wiseman, J. The great energy transition of the 21st century: The 2050 Zero-Carbon World Oration. *Energy Res. Soc. Sci.* **2018**, *35*, 227–232. [CrossRef]
7. Sovacool, B.K. How long will it take? Conceptualization the temporal dynamics of energy transitions. *Energy Res. Soc. Sci.* **2016**, *13*, 202–215. [CrossRef]
8. German Bioeconomy Council. *Bioeconomy Policy (Part II) Synopsis of National Strategies around the World*; The Office of the Bioeconomy Council: Berlin, Germany, 2015.
9. European Commission. *Innovating for Sustainable Growth: A Bioeconomy for Europe*; European Commission: Brussels, Belgium, 2012.
10. Jefferson, M. Safegurading rural landscapes in the new era of energy transition to a low carbon future. *Energy Res. Soc. Sci.* **2018**, *37*, 191–197. [CrossRef]
11. Dietz, T.; Börner, J.; Förster, J.J.; Braun, J. Von Governance of the Bioeconomy: A Global Comparative Study of National Bioeconomy Strategies. *Sustainability* **2018**, *10*, 3190. [CrossRef]
12. Kleinschmit, D.; Arts, B.; Giurca, A.; Mustalahti, I.; Sergent, A. Environmental concerns in political bioeconomy discourses. *Int. For. Rev.* **2017**, *19*, 41–55. [CrossRef]
13. CGDD (Comissariat Général au Développement Durable); SoeS (Service de l'Observation et des Statistiques). *Chiffres clés de l'Energie*; CGDD: Paris, France, 2009; 40p.
14. DGEMP (Direction Générale de l'Energie et des Matières Premières). *Scénario Energétiques de Référence: Rapport de Synthèse*; DGEMP: Paris, France, 2008; 20p.
15. Institut National de l'Information Géographique et Forestière. *Le mémento: Inventaire Forestier*; Ministère de la Transition Ecologique et Solidaire: Paris, France, 2017.
16. Ministère de l'Agriculture l'Agro-alimentaire et de la Forêt. *Programme National Forêt-Bois 2016–2026*; Ministère de l'Agriculture l'Agro-alimentaire et de la Forêt: Paris, France, 2016.
17. CGDD (Comissariat Général au Développement Durable); SoeS (Service de l'Observation et des Statistiques). *The National Strategy of Ecological Transition towards Sustainable Development—International comparisons*; Ministère de l'Ecologie, du Développement Durable et de l'Energie: Paris, France, 2017; 100p.
18. Pülzl, H.; Giurca, A.; Kleinschmit, D.; Arts, B.; Mustalahti, I.; Sergent, A.; Secco, L.; Pettenella, D.; Brukas, V. The role of forests in bioeconomy strategies at the domestic and EU level. In *Towards a Sustainable European Forest-Based Bioeconomy*; Hetemäki, L., Winkel, G., Leskinen, P., Colling, R., Eds.; European Forest Institute: Joensuu, Finland, 2017; pp. 36–51.
19. McCornick, K.; Kautto, N. The bioeconomy in Europe: An overview. *Sustainability* **2013**, *5*, 2589–2608. [CrossRef]
20. Straffas, L.; Gustavsson, M.; McCornick, K. Strategies and policies for the bioeconomy and bio-based economy: An analysis of official national approaches. *Sustainability* **2013**, *5*, 2751–2769. [CrossRef]
21. Debruyne, J. *La filière bois en Île de France*; AgroParisTech Forest Department: Paris, France, 2009.
22. Baral, A.; Guha, G.S. Trees for carbon sequestration or fossil fuel substitution: The issue of cost vs. carbon benefit. *Biomass Bioenergy* **2004**, *27*, 41–55. [CrossRef]
23. Löfstedt, R.E. The use of biomass energy in a regional context: The case of Växjö Energi, Sweden. *Biomass Bioenergy* **1996**, *11*, 33–42. [CrossRef]
24. Pedroli, B.; Elbersen, B.; Frederiksen, P.; Grandin, U.; Heikkilä, R.; Krogh, P.H.; Izakovičová, Z.; Johansen, A.; Meiresonne, L.; Spijker, J. Is energy cropping in Europe compatible with biodiversity?—Opportunities and threats to biodiversity from land-based production of biomass for bioenergy purposes. *Biomass Bioenergy* **2013**, *55*, 73–86. [CrossRef]
25. Caputo, J. *Sustainable Forest Biomass: Promoting Renewable Energy and Forest Stewardship*; Environmental and Energy Study Institute: Washington, DC, USA, 2009.
26. Sansilvestri, R.; Benito-Garzón, M.; Frascaria-Lacoste, N.; Cuccarollo, M.; Fernandez-Manjarres, J. Evaluating climate change adaptation pathways through capital assessment: Five case studies of forest social-ecological systems in France. *Sustain. Sci.* **2020**, *15*, 539–553. [CrossRef]
27. Ostrom, E. Polycentric systems for coping with collective action and global environmental change. *Glob. Environ. Chang.* **2010**, *20*, 550–557. [CrossRef]
28. OECD (Organization for Economic Cooperation and Development). *Small and Medium Enterprise Outlook*; OECD Publications: Paris, France, 2002.

29. Bastien, S. Observation participante ou participation observante? *Usages et justifications de la notion de participation observante en sciences sociales. Rech. Qual.* **2007**, *27*, 127–140.
30. Moeran, B. *From Participant Observation to Observant Participation: Anthropology, Fieldwork and Organizational Ethnography*; Copenhagen Business School Publications: Copenhagen, Denmark, 2007.
31. Davi, H.; Cailleret, M. Assessing drought-driven mortality trees with physiological process-based models. *Agric. For. Meteorol.* **2017**, *232*, 279–290.
32. Cailleret, M.; Nourtier, M.; Amm, A.; Durand-Gillmann, M.; Davi, H. Drought-induced decline and mortality of silver fir and differ among three sites in Southern France. *Annu. For. Sci.* **2013**, *71*, 643–657. [CrossRef]
33. Carnicer, J.; Coll, M.; Ninverola, M.; Pons, X.; Sanchez, G.; Penuelas, J. Widespread crown condition decline, food web disruption, and amplified tree mortality with inceased climate change type drought. *Proc. Natl. Acad. Sci. USA* **2011**, *108*, 1474–1478. [CrossRef]
34. Observatoire régional de la forêt méditerranéenne. *Données et Chiffres-clés de la Forêt Méditerranéenne*; Observatoire régional de la forêt méditerranéenne: Marseille, France, 2014.
35. Moon, K.; Blackman, D. A Guide to Understanding Social Science Research for Natural Scientists. *Conserv. Biol.* **2014**, *28*, 1167–1177. [CrossRef]
36. Gomm, R.; Hammersley, M.; Foster, P. *Case Study Method*; Sage Publications: London, UK, 2000.
37. Fabricius, C.; Scholes, R.; Cundill, G. Mobilizing Knowledge for Integrated Ecosystem Assessments. In *Bridging Scales and Knowledge Systems: Concepts and Applications in Ecosystem Assessment*; Reid, W.V., Berkes, F., Wilbanks, T., Capistrano, D., Eds.; Island Press: Washington, DC, USA, 2006; pp. 165–182. ISBN 1597260371.
38. Geaves, L.H.; Penning-Rowsell, E.C. Flood risk management as a public or a private good, and the implications for stakeholder engagement. *Environ. Sci. Policy.* **2016**, *55*, 281–291. [CrossRef]
39. Araral, E. Ostrom, Hardin and the Commons: A Critical Appreciation and a Revisionist View. *Environ. Sci. Policy* **2014**, *36*, 11–23. [CrossRef]
40. Perman, R.; Ma, Y.; McGilvary, J.; Common, M. *Natural Resource and Environmental Economics*, 3rd ed.; Pearson Education Limited: London, UK, 2003.
41. Beggs, J. The Four Different Types of Goods. Available online: https://www.thoughtco.com/excludability-and-rivalry-in-consumption-1147876 (accessed on 24 June 2021).
42. Hodge, D.; Brukas, V.; Giurca, A. Forests in a bioeconomy: Bridge, boundary or divide? *Scand. J. For. Res.* **2017**, *32*, 582–587. [CrossRef]
43. Fernandes, P.M.; Davies, G.M.; Ascoli, D.; Fernández, C.; Moreira, F.; Rigolot, E.; Stoof, C.R.; Vega, J.A.; Molina, D. Prescribed burning in Southern Europe: Developing fire management in a dynamic landscape. *Front. Ecol. Environ.* **2013**, *11*, 4–14. [CrossRef]
44. Moreira, F.; Russo, D. Modelling the impact of agricultural abandonment and wildfires on vertebrate diversity in Mediterranean Europe. *Landsc. Ecol.* **2007**, *22*, 1461–1476. [CrossRef]
45. Moreira, F.; Viedma, O.; Arianoutsou, M.; Curt, T.; Koutsias, N.; Rigolot, E.; Barbati, A.; Corona, P.; Vaz, P.; Xanthopoulosh, G.; et al. Landscape–wildfire interactions in southern Europe: Implications for landscape management. *J. Environ. Manag.* **2011**, *92*, 2389–2402. [CrossRef] [PubMed]
46. Costanza, R. Valuing natural capital and ecosystem services toward the goals of efficiency, fairness, and sustainability. *Ecosyst. Serv.* **2020**, *43*, 101096. [CrossRef]
47. Schultz, L.; Folke, C.; Österblom, H.; Olsson, P. Adaptive governance, ecosystem management, and natural capital. *Proc. Natl. Acad. Sci. USA* **2015**, *112*, 7369–7374. [CrossRef]
48. Cardinale, B.J.; Matulich, K.L.; Hooper, D.U.; Byrnes, J.E.; Duffy, E.; Gamfeldt, L.; Balvanera, P.; O'Connor, M.I.; Gonzalez, A. The functional role of producer diversity in ecosystems. *Am. J. Bot.* **2011**, *98*, 572–592. [CrossRef]
49. Magnan, A. De la vulnérabilité à l'adaptation au changement climatique: Éléments de réflexion pour les sciences sociales. In *Risques Côtiers et Adaptations des Sociétés*; ISTE Group: London, UK, 2014; pp. 241–274.
50. Ravera, F.; Tarrason, D.; Simelton, E. Envisioning adaptive strategies to change: Participatory scenarios for agropastoral semiarid systems in Nicaragua. *Ecol. Soc.* **2011**, *16*, art20. [CrossRef]
51. Olsson, P.; Folke, C.; Hahn, T. Social-Ecological Transformation for Ecosystem Management: The Development of Adaptive Co-management of a Wetland Landscape in Southern Sweden. *Ecol. Soc.* **2004**, *9*, 2. [CrossRef]
52. Rifkin, J. *The Third Industrial Revolution. How Lateral Power Is Transforming Energy, the Economy, and the World*; MacMillan, P., Ed.; St. Martin's Publishing Group: Manhattan, NY, USA, 2011; 304p.
53. Isnard-Dupuy, P. Face au chantage aux subventions les parcs régionaux acceptent la centrale de Gardanne. *Reporterre Newspaper*, 4 October 2017.
54. Upreti, B.R.; Van Der Horst, D. National renewable energy policy and local opposition in the UK: The failed development of a biomass electricity plant. *Biomass Bioenergy* **2004**, *26*, 61–69. [CrossRef]
55. Jenkins, K.; Sovacool, B.K.; Mccauley, D. Humanizing sociotechnical transitions through energy justice: An ethical framework for global transformative change. *Energy Policy* **2018**, *117*, 66–74. [CrossRef]
56. Isnard-Dupuy, P. L'ONF encourage la biomasse industrielle au detriment des forêts et du climat. *Reporterre Newspaper*, May 2018.

57. Pulighe, G.; Bonati, G.; Colangeli, M.; Morese, M.M.; Traverso, L.; Lupia, F.; Khawaja, C.; Janssen, R.; Fava, F. Ongoing and Emerging Issues for Sustainable Bioenergy Production on Marginal Lands in the Mediterranean region. *Renew. Sustain. Energy Rev.* **2019**, *103*, 58–70. [CrossRef]
58. Benkler, Y. *The Penguin and the Leviathan: How Cooperation Triumphs over Self-Interest*, 1st ed.; Crown Pubishing Group: New York, NY, USA, 2011; ISBN 978-0-385-52576-3.
59. Uehara, T.; Hidaka, T.; Tsuge, T.; Sakurai, R.; Cordier, M. An adaptive social-ecological system management matrix for guiding ecosystem service improvements. *Ecosyst. Serv.* **2021**, *50*, 101312. [CrossRef]
60. Uehara, T.; Hidaka, T.; Matsuda, O.; Sakurai, R.; Yanagi, T.; Yoshioka, T. Satoumi: Re-connecting people to nature for sustainable use and conservation of coastal zones. *People Nat.* **2019**, *1*, 435–441. [CrossRef]
61. Jones, N.A.; Shaw, S.; Ross, H.; Witt, K.; Pinner, B. The study of human values in understanding and managing social-ecological systems. *Ecol. Soc.* **2016**, *21*, 15. [CrossRef]
62. Elyakime, B.; Cabanettes, A. Forest Policy and Economics How to improve the marketing of timber in France? *For. Policy Econ.* **2009**, *11*, 169–173. [CrossRef]
63. Aggeri, F.; Godard, O. Les entreprises et le développement durable. *Entrep. Hist.* **2006**, *45*, 6–19. [CrossRef]

forests

MDPI

Article

Sustainable Forest Management Evaluation Using Carbon Credits: From Production to Environmental Forests

Noriko Akita [1,*] and Yasuo Ohe [2]

[1] Graduate School of Horticulture, Chiba University, Chiba 271-8510, Japan
[2] Faculty of International Agriculture and Food Studies, Tokyo University of Agriculture, Tokyo 156-8502, Japan; yo207358@nodai.ac.jp
* Correspondence: noriko@faculty.chiba-u.jp; Tel.: +81-47-308-8931

Abstract: The biodiversity and carbon dioxide absorption function of forests have received attention due to global warming. However, most of the world's forests are general production forests. Since production forests are maintained by production activities, a decrease in production or abandonment of management leads to a decline in forest functions and increases the risk of disasters such as landslides. Against this background, the retention approach has been proposed as a way to convert general production forests into forests with enhanced environmental functions, but it has rarely been applied due to technical and cost barriers. This study focuses on cost barriers and examines the possibility of introducing a retention approach to converting production forests to environmental forests, using Japan as a case study. About 70% of Japan's land area is covered with forests, 40% of which are production forests. However, due to the sharp decline in demand for timber in recent years and price competition with imported timber, the selling price of timber has fallen below the cost of managing production forests, and the management of many production forests has been abandoned. The dilemma is that the retention approach applied to the wood production process cannot be applied to forests where production activities are stagnant. Therefore, we explored the possibility of recovering the necessary costs with carbon credits that are available in the Japanese market. We calculated the cumulative carbon stocks of carbon dioxide in production forests by age, using intensity, and estimated how many years after planting the combined costs of normal production forests management and the retention approach would balance out. Our calculations show that even if carbon credits were sold at the lowest market price, the balance of payments would be balanced about 30 years after planting, resulting in a net profit from the sale of the wood.

Keywords: forest sustainability; production forests; environment forests; carbon credit; forest management; retention approach

Citation: Akita, N.; Ohe, Y. Sustainable Forest Management Evaluation Using Carbon Credits: From Production to Environmental Forests. *Forests* **2021**, *12*, 1016. https://doi.org/10.3390/f12081016

Academic Editors: Juan F. Fernández-Manjarrés and Roxane Sansilvestri

Received: 26 June 2021
Accepted: 23 July 2021
Published: 30 July 2021

1. Introduction

Forests play a critical multifunctional role in environmental conservation. Significant and irreplaceable forest functions include ecological aspects such as the preservation of ecosystems and biodiversity, and the carbon-dioxide absorption that mitigates global warming [1,2]. In particular, the absorption of carbon dioxide is essential for controlling the destruction of the ecosystem itself as a result of global warming. Currently, there are only two ways available to reduce atmospheric carbon dioxide: either to reduce or absorb emissions.

Regarding forest functions, forests such as tropical rainforests with special ecological rarity have historically attracted substantial attention [3,4]; however, Lindenmayer et al. [5] highlighted that 11% of the world's forests are protected forests and 4% are intensive wood production forests, while the remaining 85% are neither, nor emphasized the importance of focusing on these. The retention approach has been proposed as a forest management method for improving ecosystem quality in such 'general' production forests [5]. Data from relevant research has led to the retention forest approach being globally recognized in recent times as an effective means for converting production forests into environmental forests [6–9].

Despite the consensus on the importance of the retention forest approach, only a few countries have adopted it [10] due to the technical issues and cost barriers. From a technical point of view, research is still necessary to assess which logging methods can be beneficial for a variety of ecosystems [10]. In terms of cost, there has been insufficient discussion on how to deal with the additional costs of adopting a retention approach. Above all, as the selling price of timber is being reduced due to intensifying international competition, it is becoming difficult to secure even the normal maintenance cost of production forests when the production forests are located on steep slopes with low productivity, as is the case in Japan.

Long-term forest management planning and careful harvesting operations are essential in implementing the retention approach, which require additional costs covering detailed forest conditions surveys and biodiversity-based surveys ensuring proper logging planning. Barreto et al. [11] estimated that approximately US 72/ha more are required to develop a forest logging management plan, while having no plan is more profitable in the short term. Arnott and Beese [12] showed that harvesting costs increase by approximately 50% when biodiversity-friendly logging is carried out. The total cost required not only for felling but also shipping as timber will increase by at least 10% [10]. However, as mentioned earlier, it is almost impossible to meet these costs in production forests with low productivity. In the case of production forests that have been abandoned, there is no opportunity to adopt such a retention approach. In other words, a social dilemma arises in which the production forests with the greatest need for conversion to environmental forests do not have the opportunity to do so because they cannot bear the cost of conversion.

To cope with the costs of converting to environmental forests, we focused on carbon credits, which are generally used as payments for contributions to the environment. This is because the forestry industry in many countries, including Japan, is already supported by subsidies, and if the dependence on subsidies increases, the forestry industry itself will be weakened. In Japan, even the number of forestry workers who receive subsidies is already decreasing. Therefore, we examined the possibility of using marketable carbon credits as a way to establish sustainable forestry in a market economy. There are already cases where carbon dioxide fixation by forest management is incorporated into the carbon offset mechanism. At present, however, carbon offsetting tends to be biased toward renewable energy, and carbon offsetting for forest management is not widely used. This study uses Japan as its case study, where the steep terrain prevents efficient forestry, and the management of many production forests has been abandoned. Japan has a system of carbon offsetting, but it was only used once in 2015, and since then, carbon offsetting has not been used for forest management [13]. One of the reasons why carbon credits have not been actively used for forest management is that it is difficult to evaluate carbon credits for forest management [14,15].

Malmsheimer et al. [15] highlighted the following three issues regarding carbon credit for forests: first, it is difficult to calculate costs other than for new tree-planting projects; second, the carbon absorption estimation baseline has not yet been clearly established; and third, it is difficult to incorporate carbon credits accumulated in the past. However, we developed a simple model of forest carbon sequestration and forest management costs in order to understand whether production forests that cannot afford management costs can be converted to environmental forests using the retention approach. The reason for this is that we thought that it would be best to examine the possibility here first, and then perform detailed calculations. The model is based on the amount of carbon dioxide absorbed in production forests by different age groups. The amount of carbon dioxide absorbed is valued as carbon credits that can be traded in the market.

We hypothesize that carbon credits can cover the cost of the retention approach and proper management of production forests; this hypothesis we shall henceforth refer to as the 'market carbon trade hypothesis'. The amount paid to the production forests is determined by the amount of carbon accumulated in it. We aim to elucidate under which basic conditions the hypothesis is valid.

We selected Japan as the target location for this "market carbon trading hypothesis". There are two reasons for this. First, the demand for timber in Japan has plummeted over the past 50 years, leaving production forests unmanaged and unattended. The government has intervened many times with subsidies to the forestry industry, but without significant effect. Timber is exposed to international competition, and in Japan, where production forests are located in steep mountains, the selling price of timber is less than the cost of managing the production forests. Although 70% of the country's land area is covered with forests, and about 40% of these are production forests, the production forests are left unattended, which poses a challenge to the conservation of ecosystems and risks inducing disasters. There is a need to examine whether carbon credits can overcome the dilemma that the more production forests that need to be converted to environmental forests, the more difficult it is to introduce a retention approach.

Second, carbon credits are traded in the Japanese market; therefore, it is possible to evaluate the current carbon price. For the payment of the environmental conservation function of forests, some countries and regions have already adopted mandatory or voluntary emissions-reduction mechanisms [16,17]. Preece et al. [18] showed that the advanced carbon farming system established in Australia in 2011 plays a key role in sustainable forest management. It is important to examine the possibilities in countries other than those already studied, and the results obtained from the study can provide useful suggestions internationally.

To examine the above hypothesis, we adopt the following three procedures: First, we estimate the carbon absorption from production forests using data of the Japanese forest research institutes (Forestry and Forest Products Research Institute, Japan; FFPRI), thereby revealing the carbon baseline setting, which is important when introducing the carbon credit system to forests. Second, we calculate the cost of proper forest management. We use the management cost calculated using the forestry association data on the proper management cost of production forests. We assume that the retention approach will be adopted during logging and forest management. The cost of adopting the retention approach is based on earlier research, which assumes that the cost of logging will increase by 10% [10]. The cost of planning and investigating the retention approach was excluded from this calculation because it can be formulated separately from forest management activities. Third, we simulate the price of accumulated carbon in production forests by using multiple carbon-credit prices traded on the market. Based on this simulation, we judge the price and period for which carbon credits balance management costs. We infer that if the valuation of carbon credits exceeds the cost required for proper management of production forests and adaption of the retention approach, it is possible to apply the retention approach. We propose to solve the dilemma of degraded production forests by converting production forests into environmental forests through the introduction of carbon credits in forests where productivity cannot be increased. Finally, we present the policy implications of this study.

2. Materials and Methods

2.1. Literature Review

We reviewed production forests on a global scale and addressed the retention approach, a proposed new management method for production forests. We elaborate on the evaluation of the function and amount of carbon absorption in forests and discuss the contribution of carbon credits related to environmental issues.

We provide an overview of the current state of forests globally, and then focus on production forests. The total forest area globally is 4.60 billion hectares (ha), accounting for about 31% of the world's land-surface area [19,20]. The global forest area has been generally in decline; however, it has recently slightly increased due to continuous efforts to maintain and preserve the diverse functions of forests. Nevertheless, Butchart et al. [21] revealed that biodiversity loss remains severe, despite efforts to establish sustainable forest management in some areas. Researchers highlight that future global climate change could affect forest ecosystems [22,23]. In this regard, the 2015 United Nations Sustainable Development Summit set 17 Sustainable Development Goals (SDGs). The fifteenth SDG states: "Protect, restore and promote sustainable use of terrestrial ecosystems, sustainably manage forests, combat desertification, and halt and reverse land degradation and halt biodiversity loss" [24].

Particularly, critical functions that forests have for the environment are ecological aspects such as the preservation of ecosystems and biodiversity, and the carbon-dioxide absorption function that mitigates global warming [25]. Lindenmayer et al. [5] found that most forests globally are neither formally protected nor dedicated to intensive wood production. The retention approach has been proposed as a forest management method for improving the quality of ecosystems in such general production forests [5]. Groot et al. [26] showed that transforming Canadian traditional forests into retention forests is the most practical management method of present forestry. It has been shown that long-term forest planning is required to achieve retention forestry. Gustafsson et al. [6] evaluated the retention approach as a scientifically validated approach that can resolve conflicting goals regarding timber production and biodiversity conservation and preserve a degree of species richness equivalent to that of primary forests. Currently, the retention forestry approach is recognized globally as a conservation tool in production forests [9]; however, its implementation is limited to some countries and regions due to technical and cost barriers [8]. Additionally, research on logging methods that contribute to diverse biodiversity are ongoing [10].

We elaborate on the evaluation methods of the function and amount of carbon absorption in forests. Relevant studies can be grouped into three types: estimates based on remote sensing, estimates based on actual measurements, and calculation of carbon content using equations and models. Research with estimates based on remote sensing aims mainly to assess a meta-level carbon storage over wide forested areas. Such research is usually conducted on tropical rainforests, where the impact of area reduction on the environment is significant [27,28]. Saatchi et al. [4] mapped the total carbon stock of living biomass above and below ground, using a combination of data from in situ inventory plots and satellite light detection and ranging (Lidar) samples of forest structure to estimate carbon storage. They created a benchmark map of the forest carbon stock in the early 2000s. Baccini et al. [3] used multi-sensor satellite data to estimate aboveground carbon density of living woody vegetation in pan-tropical ecosystems and its spatial distribution. They highlighted that one of the reasons for adopting this method was to reduce the cost of assessing the carbon stock in forests [4]. Actual measurements of carbon accumulation in forests capture the changes in carbon content due to logging. Rainforests are often selected for investigations [29]. Keith et al. [30] studied the Australian temperate moist eucalyptus forests and some other types of forest carbon stocks. Such studies provide important basic data for carbon calculations; however, the areas of application are limited since logging is a requirement. In addition, since the amount of accumulated carbon in a forest varies greatly depending on the tree species, it is difficult to evaluate the entire amount without investigating the tree species [31].

The calculation of carbon content by equations and models is used to obtain a rough amount of the carbon storage of the entire forest. Diverse international and national researchers and research institutes began examining the carbon-absorption function of forests and assessing carbon dioxide reductions based on the 1997 Kyoto Protocol by the United Nations Framework Convention on Climate Change (UNFCCC). Chave et al. [32] developed a regression model involving only wood density and stem diameter to estimate the amount of tropical biomass and the contributions of the tropical forest biome and deforestation to the global carbon cycle. Jenkins et al. [33] compiled diameter-based allometric regression equations for estimating the total aboveground and component biomass, defined in dry-weight terms, for trees in the United States. These earlier studies are mainly used to create reference values and obtain a rough idea of changes in carbon content. Since the purpose of this study is also to test the hypothesis, we decided to use the basic unit calculated by the Forestry and Forest Products Research Institute, Japan [34].

We also focused on empirical studies on forest management costs and carbon credit contributions to the environment. Canadell and Raupach [35] claimed that forests absorb billions of tons of CO_2 globally every year; an economic subsidy of hundreds of billions of dollars would be necessary in order to create an equivalent sink. In some countries and regions, the payment for the environmental conservation function of forests have already adopted mandatory or voluntary emissions-reduction mechanisms [16,36]. Preece et al. [18] highlighted that Australia's 2011 Carbon Farming Initiative incorporates carbon credits, supports reforestation, and has been beneficial for sustainable forest management. Kurz et al. [37] proposed a carbon-dynamics model for forestry and land use change for Canada. Morse et al. [38] analyzed the effectiveness of forestry legislation in Costa Rica, which introduced payments for environmental services in 1996. Results showed that the payment of environmental services increased forestry retention, and the carbon stocks of secondary forests approached the levels of primary forests after 25–30 years. Kayo et al. [39] evaluated the possibility of converting production to environmental forests in Japan, similar to this study; however, carbon credits had not been introduced at that point, and the validity of the study was not assessed. Ellison et al. [40] claimed that current approaches of resource-based carbon accounting consider only a fraction of the forest's potential. Currently, Japan is trading in the carbon credit market, and it is possible to explore its potential.

We acknowledge these studies and will consider the possibility of utilizing carbon credits for the transition from production to environmental forests. Evaluating the distribution of credits instead of mandatory taxes could enable private initiatives to contribute to environmental conservation without increasing the burden on citizens. If attempts to revitalize forestry with subsidies are not very effective, as in the case with Japan, it becomes necessary to introduce alternative mechanisms as well. Carbon credits through the market can stimulate proper forest management. Furthermore, the introduction of the retention approach can enhance multifaceted forest functions, such as biodiversity, thereby creating value beyond carbon and increasing the interest in forest management. Most Japanese production forests in the case-study area have only cedar tree species [41], so it is possible to estimate the amount of carbon reserves more accurately. The 'market carbon trade hypothesis' promotes the reduction of the burden related to the conversion of production to environmental forests by appropriately evaluating their carbon absorption function. Conversion of production to environmental forests might occur in many regions in the future; our findings provide meaningful insight into the sustainable management and ecological-value increase of production forests. Our results will be useful to stakeholders of production forests with decreasing demand and of the global environment.

2.2. Framework for the Research Design

The analysis consists of two parts. The first part analyzes the changes in the environment surrounding production forests in Japan, the target area of the case study.

As previously mentioned, 70% of Japan's land area is covered by forests, 40% of which are production forests. However, the utilization rate of forests is less than 1%, which is the lowest in the OECD. Neglected production forests need to be converted to environmental forests in order to prevent landslides and restore ecosystems. However, the dilemma is that the retention approach is difficult to implement in planted forests that need to be converted to environmental forests because the retention approach for conversion from production forests to environmental forests assumes production activities. In this study, we propose a method of converting production forests, which are difficult to implement the retention approach through production activities, into environmental forests. In Japan's production forests, this analysis has shown that recent lifestyle changes have directly affected the forest environment. This situation of lifestyle influencing forests may not be limited to Japan. Therefore, we considered a system that would support not only the people of a particular region, but society as a whole to take responsibility for the sustainable management of forests.

The second part of the study examines the potential for carbon credits to cover the costs of managing and transitioning from production forests to environmental forests using the retention approach, mostly through simulation. The amount of carbon stored in a properly managed plantation forest by age class is simulated and evaluated with several carbon credit prices available in the market. We will identify the market price and time period required to adopt a retention approach over and above the management costs of production forests.

2.3. *Simulation Framework for Evaluating the Carbon Absorption Function of Production Forests*

2.3.1. Basic Unit of Carbon Absorption

We employed the definition of the carbon dioxide absorption model of the FFPRI in Japan [34]. In accordance with the Kyoto Protocol, FFPRI conducted an analysis of the carbon-absorption function in typical Japanese forests. Our simulation targets Japanese cedar, which is a typical Japanese production forest tree species. Japanese cedar, cypress, and broad-leaved trees are typical tree species in Japanese production forests, and among them, Japanese Cedar has the highest proportion.

Figure 1 shows the basic numerical value of carbon absorption per hectare of a general Japanese cedar production forest. The cumulative carbon absorption (t-C/ha) is based on the FFPRI model. Figure 1 shows the cumulative carbon absorption of cedar by age and the carbon absorption period. From about the age of ten years, the growth rate of cedar gradually increases, and the absorption of carbon dioxide also increases sharply. The reason why the carbon absorption per period seems to decrease significantly in the 25th year is because thinning takes place during this period, reducing the number of trees present per hectare. Beyond the age of 50 years (which corresponds to the main logging season), the growth rate slows and the rate of increase in carbon dioxide absorption also slows; however, the cumulative carbon absorption of cedar production forests continues to increase. Figure 1 shows the weight of carbon, and in order to convert this to the weight of carbon dioxide, we multiply it by 44/12. The molecular weight of CO_2 is 44 (the atomic weight of C is 12 and the atomic weight of O is 16, so $12 + 16 \times 2 = 44$). Since the atomic weight of C is 12, we can multiply the carbon equivalent weight by 44/12 to get the carbon dioxide equivalent. So 44t-CO_2 is equal to 12t-C.

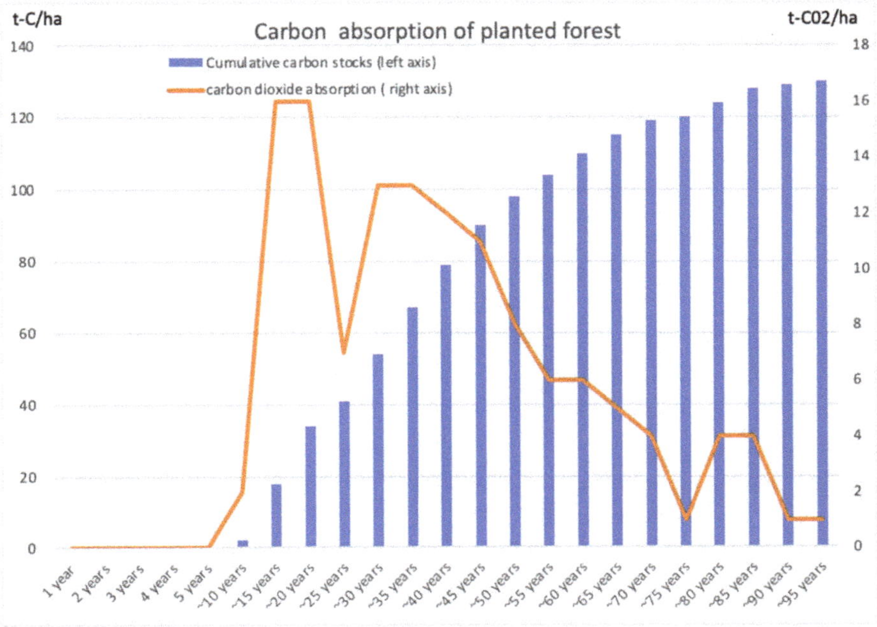

Figure 1. Cumulative carbon absorption and carbon-absorption period in a general Japanese cedar production forest. Source: [34].

2.3.2. Management Costs of Production Forests

For the annual management costs of production forests, we used the values published by the Ministry of Agriculture, Forestry, and Fisheries [42]. The name of the data is "Forestry Management Statistics Survey", and is published as official statistical data of the government. Since the annual management costs after the 51st year, which is the main harvesting period, were not provided, we assumed they were equivalent to the management costs required in years 46 to 50.

The retention approach requires selective logging. Although previous studies have shown that the additional cost of adopting a retention approach is required at the time of harvesting and does not add to normal management costs, leaving dead trees in the forest will increase the amount of management effort. There are various estimates of the cost of the retention approach, and it is currently not fully determined. Then, this study assumed a 10% increase in logging costs when adopting a retention approach based on previous studies. In addition, we have added 10% to the cost required for management. These transactions are conducted in Japanese Yen; however, we have converted them to US dollars for the sake of the readers' understanding. The conversion rate from Japanese yen to US dollar used is JPY 110, i.e., the average of the Telegraphic Transfer Selling Rate (TTS) in 2019. These figures are shown in Table 1.

Table 1. Maintenance costs of production forests in Japan (Unit: ha). Source: [42].

Period	Years	Type of Maintenance Activity	Normal Costs (Yen/ha)	Normal Costs (USD/ha)	Cost with Retention Approach (USD/ha)
Forestation period	1	Afforestation	418,679	3806	-
	2	Cutting underbrush	232,502	2114	-
	3		119,760	1089	-
	4		163,112	1482	-
	5		9282	844	-
Growth period	6–10	Thinning	47,738	434	477
	11–15		21,079	192	211
	16–20		17,195	156	172
	21–25		38,195	347	382
	26–30		21,364	194	214
	31–35		12,466	113	125
	36–40		9734	88	97
	41–45		6468	59	65
	46–50		9862	90	99
Maturity period	51–	Maintenance	9862	90	99

One of the costs of implementing the retention approach is the cost of research and planning, but since the retention approach is still in the experimental stage, the cost has not been presented in previous studies. We assumed that the cost of such planning would be borne by the national and local governments, not the forestry community. This is because the scale of Japan's forestry industry is not large, and the work involved in forestry itself is already supported by subsidies. For this reason, this study does not take into account the research and planning costs that would be required to introduce a retention approach. The purpose of this study is to focus on the costs of forest management if a retention approach is introduced, and to examine the possibility that this could be covered by the market.

2.3.3. Setting Carbon Credit Prices in the Simulation

Under the United Nations Framework Convention on Climate Change (UNFCCC), each country has declared carbon reductions, and Japan has set a high target of 26% reduction by 2030 compared to 2013 [43]. A key for achieving this goal is reducing emissions, especially in large cities with high carbon footprints. Tokyo is the largest city in Japan, where companies are concentrated, and the Tokyo Metropolitan Assembly passed a revision of the Ordinance on the Environment to Ensure the Health and Safety of Citizens in 2008, mandating the reduction of greenhouse gas emissions from large-scale business establishments (i.e., those that used annually more than 1500 kL of crude oil equivalent in terms of fuel, heat, and electricity). Based on this, the Tokyo Metropolitan Government introduced a cap-and-trade system to reduce carbon dioxide emissions from 2010 onward. This is the first urban-area cap-and-trade system in the world that covers office buildings, and the Tokyo Metropolitan Government has set its own reduction targets: by 8% between 2010 and 2014, and by 17% between 2015 and 2019. As a means of achieving this goal, the Tokyo Metropolitan Government has set up a carbon credit trading system [44,45].

There are five kinds of credits for emissions trading. The first is excess reduction credit, the second is small and medium-sized enterprise credit, the third is large enterprise credit outside Tokyo, the fourth is neighboring prefecture credit of Tokyo, and the fifth is renewable energy credit. Only two types of valuation–price surveys are being conducted: renewable energy and excess reduction credits. Since excess abatement credits are specific to transactions between individual companies, this study refers to the prices of renewable energy credits that are established as independent credits. The Tokyo Metropolitan Government has published the 10-year price fluctuations of renewable energy credits from 2011 to 2020. The highest price is JPY 12,500, the lowest price is JPY 5500, the average price over 10 years is JPY 8335, and the most recent price is JPY 5600 (March 2020) [46]. We converted these to dollars using the TTS prices mentioned above. These figures are shown in Table 2. We decided to run our simulations using these four values.

Table 2. Trading price for carbon credits [1] (unit: USD/t-CO_2; based on [46]).

Type	Lowest Price	Highest Price	Average Price	Most Recent Price (March 2020)
Renewable energy credit	JPY 5500 USD 50	JPY 12,500 USD 114	JPY 8335 USD 76	JPY 5600 USD 51

[1] The conversion rate from Japanese Yen to US dollars used is JPY 110, by TTS in 2019.

3. Results

3.1. Challenges in Sustainable Management of Japanese Forests

3.1.1. Changes in the Forestry and Lumber Industries in Japan

Japan is known as the "Land of Forests" in the world. Forests are deeply involved in the lives and culture of the Japanese people, and the term "satoyama" (sato; human habitation, yama; mountain) was born in Japan [47]. Japan's forests underwent major changes after World War II in 1945. As the population grew after the war, demand for lumber surged in Japan, where wooden houses were the standard form of housing. The lumber-producing regions could not keep up with production and began to rely partially on imports. On the other hand, forest owners, attracted by the high price of lumber, enthusiastically engaged in afforestation activities, focusing on cedar, which is used for building materials. As a result, as of 2012, 41% of Japan's total forest area is production forests, of which 44% is made up of cedar [48].

In Japan, it takes about 50 years from planting to harvesting. However, during the past 50 years, Japanese lifestyles and economic conditions have changed dramatically. In urban areas, reinforced concrete housing complexes instead of wooden houses have become the main way of living [49]. As a result, the demand for wood for housing has decreased. In addition, Japan's mountain forests have many steep slopes, and the cost of producing lumber is relatively high, making domestic lumber more expensive than imported lumber and making it uncompetitive. Currently, the cost of managing production forests exceeds the selling price of timber, and there are many production forests that are poorly managed and production activities have stagnated. Figure 2 shows the ratio of timber production to forest accumulation in the OECD (Organization for Economic Cooperation and Development) countries [50]. Japan has the lowest ratio of wood production to forest accumulation among the OECD countries, at only 0.47% in 2015. This is about one-tenth of Scandinavia, which has the highest amount of wood production relative to forest accumulation and is the lowest. Table 3 shows this in figures. Japan's wood production to forest accumulation was only 39.17% in 2015 compared to the average of OECD countries.

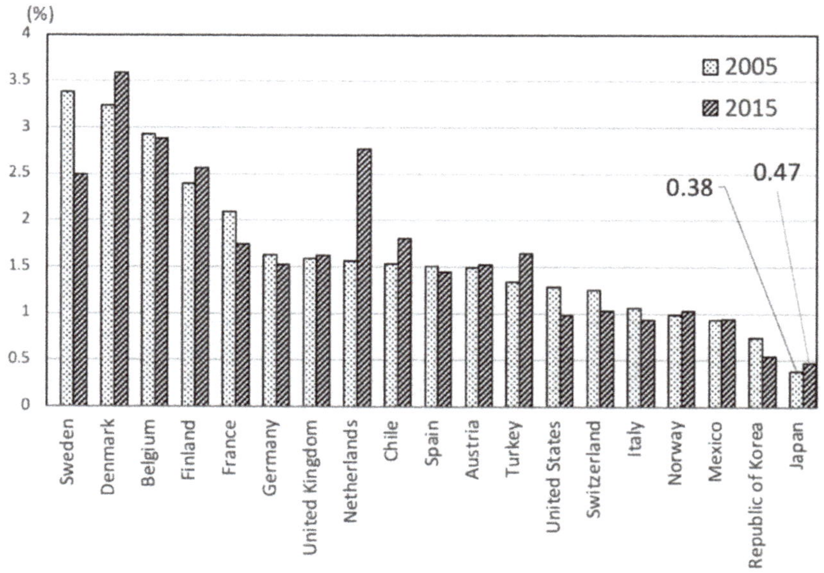

Figure 2. Ratio of timber to forest production of some OECD countries, based on [50].

Table 3. Timber to forest production ratio, based on [50].

Classification	Value of All OECD Member Countries			Value for Japan	
Year	Amount of Wood Production (million m^3)	Amount of Forest Accumulation (million m^3)	Wood Production/ Accumulation (%)	Wood Production/ Accumulation (%)	Comparison with OECD Average (%)
2005	1046	76,529	1.37	0.38	27.74
2015	1022	85,180	1.20	0.47	39.17

In light of this situation, it must be said that it is currently difficult to convert production forests into environmental forests through use. In order to properly manage abandoned production forests and convert them into environmental forests, a new added value that complements the use-value of wood is necessary. This study explores the possibility of evaluating the carbon sequestration function of production forests as carbon credits to see if it is possible to cover the costs of managing production forests and renewing them into environmental forests.

3.1.2. Forest Age Structure

In Japanese forestry (practice), forests are classified into age classes that have 5 years width. The highest possible age class is 20 with a final age of 100 years. Figure 3 shows the age classes of Japan's production forests in 2017. The area occupied by the 10th–12th age classes is the largest among the production forests, indicating that afforestation was actively carried out in the late 1960s and 1970s. Today, 50 to 60 years have passed since the time of planting, and most of the trees are in a condition suitable for logging.

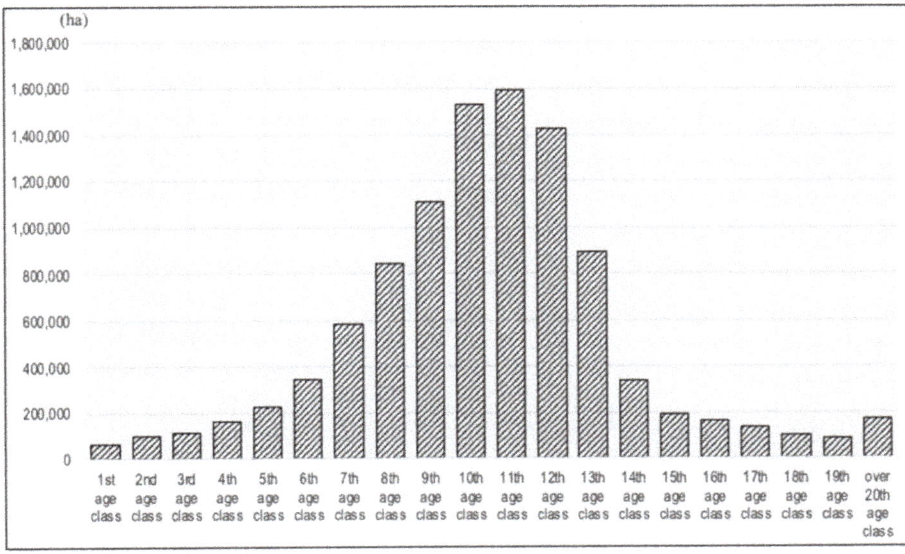

Figure 3. Total area of production forests in Japan presented by age classes in 2017, based on [51].

In the 1960s, when vigorous afforestation was carried out, Japan was in a period of rapid economic growth, with a rapidly growing population and increasing demand for housing. However, due to the subsequent spread of reinforced concrete housing complexes and price competition with imported lumber, the domestic demand for lumber rapidly declined. As a result, the area of forestation has also been decreasing since its peak around 50 years ago. In addition, even when the timber has reached the optimum age for harvesting, the cost of managing production forests is higher than the profits from timber sales, so production forests are left unmanaged.

3.1.3. Forest Management

The main activities in Japan's production forests can be classified into four processes: logging, thinning, underbrush clearing, and afforestation. After planting, weeding is carried out for 20 years, and from the 20th year to the 50th year, thinning is performed to allow the tree trunks to grow. About 50 years after planting, the trees in the production forests grow to a size suitable for harvesting. In Japan, there are about 10 million hectares of production forests [52], of which 980,000 hectares, or about 10%, are outsourced to forest cooperatives or private companies for management [53]. Table 4 shows the area and ratio of logging, thinning, underbrush cutting, and afforestation by contracted forestry companies on these 980,000 ha. Figure 3 shows that although it is clear that most of Japan's production forests have reached the appropriate age for logging, only 4.47% of the contracted area is logged, and only 22.01% of the area is thinned. Furthermore, it is a fact that the area entrusted for the project is 10% of the production forests. This data shows that it is impossible to change the forest structure through forestry.

Table 4. Ratio of management work to planted forest area in 2015, based on [52]. (Unit: ha)

Logging	Thinning	Cutting Underbrush	Afforestation
43,825	215,771	148,833	24,401
4.47%	22.01%	15.19%	2.49%

The data in Table 4 also shows another problem: in 2015, only about half of the production forests area that was cut down was planted. If this trend continues, tens of thousands of hectares of forest area will be lost every year. Due to this situation, the law was amended in 2017 to require afforestation after logging. However, since afforestation is costly, the dilemma arises that making afforestation mandatory will further hinder the progress of logging. In order to turn neglected production forests into environmental forests, we need an engine to overcome this dilemma.

3.2. Verification of the Market Carbon Trade Hypothesis

3.2.1. Calculation of Carbon Absorption and Management Cost by Age

We performed a carbon credit assessment of the carbon dioxide absorption of forests. Table 5 shows the amount of carbon absorbed and the management cost by age. In Japan, trees in planted forests are divided into the following three periods according to the degree of their growth: until the 5th year after planting is the "forestation period"; from the 6th to 50th year after planting is the "growth period"; and from the 51st to 95th year after planting is the "maturity period". The amount of carbon absorbed in each period was calculated based on the criteria in Section 3.2.1. The carbon absorption during the forestation period was zero. At the end of the forestation and the beginning of the growing period, the trees grew large, and the amount of carbon absorbed increased rapidly. In Japan, the cumulative amount of carbon absorbed in forests is published in five-year groups; hence, Table 5 shows the amount of carbon absorbed in each period.

Table 5. Calculation of carbon absorption and management cost by age.

Period	Managing Activities	Years	Accumulated Carbon Amount		Carbon Dioxide Absorption during the Period (t-CO^2 ha^{-1})	Management Cost (USD)	Including Retention Approach Cost (USD)
			Cumulative Carbon Content (t-C ha^{-1})	Periodic Carbon Accumulation (t-C ha^{-1})			
Forestation period	Afforestation	1	0	0	0.00	3806	3806
	Cutting underbrush	2	0	0	0.00	2114	2114
		3	0	0	0.00	1089	1089
		4	0	0	0.00	1483	1483
		5	0	0	0.00	844	844
Growth period	Thinning	6–10	2	2	7.34	434	477
		11–15	18	16	58.72	192	211
		16–20	34	16	58.72	156	172
		21–25	41	7	25.69	347	382
		26–30	54	13	47.71	194	213
		31–35	67	13	47.71	113	124
		36–40	79	12	44.04	88	97
		41–45	90	11	40.37	59	65
	Maintenance	46–50	98	8	29.36	90	99
Maturity period	Maintenance	51–55	104	6	22.02	90	99
		56–60	110	6	22.02	90	99
		61–65	115	5	18.35	90	99
		66–70	119	4	14.68	90	99
		71–75	120	1	3.67	90	99
		76–80	124	4	14.68	90	99
		81–85	128	4	14.68	90	99
		86–90	129	1	3.67	90	99
		91–95	130	1	3.67	90	99

We then calculated the management costs required for each period. This was based on Section 3.2.2 and Table 1 and was calculated assuming that the retention approach was implemented; since the retention approach involves logging operations, the cost was added after the sixth year, when thinning begins. The calculation assumes a 10% increase in management costs from previous studies.

3.2.2. Evaluation of Pay Ability by Carbon Credit

Finally, we assessed carbon credits against the carbon sequestration of our production forests. Table 6 shows the carbon sequestration and management costs of production forests by forest age, and the price at which carbon credits would be balanced against this. The prices of carbon credits were estimated as shown in Table 2, with the lowest price of USD 50, the highest price of USD 114, the average price of USD 76, and the latest price of USD 51. In the first five years of the planting period, carbon sequestration is counted as zero. Only administrative costs are incurred, and the balance is always negative. However, from the sixth year, the carbon sink of the production forests rises rapidly, and even if the carbon credit is the lowest at USD 50, the price of the carbon credit balances the price needed to cover the cost after 30 years. After 50 years, when the trees are ready to be harvested, the carbon sink will slow down, but even if we adopt the retention approach, we will still be able to make a positive balance until the end.

Table 6. Simulation of cedar production-forests management costs per ha and renewable credit pricing (US dollars) [1].

Period	Years	(a) [2]	(b) [3]	Valuation Price of Carbon Credits				Balance of Payments			
				50 USD [4]	51 USD [5]	76 USD [6]	114 USD [7]	50 USD [4]	51 USD [5]	76 USD [6]	114 USD [7]
Forestation period	1	0.00	3806	0	0	0	0	−3806	−3806	−3806	−3806
	2	0.00	2114	0	0	0	0	−5920	−5920	−5920	−5920
	3	0.00	1089	0	0	0	0	−7009	−7009	−7009	−7009
	4	0.00	1482	0	0	0	0	−8491	−8491	−8491	−8491
	5	0.00	844	0	0	0	0	−9335	−9335	−9335	−9335
Growth period	6–10	7.33	477	367	374	557	836	−9446	−9439	−9255	−8977
	11–15	58.67	211	2933	2992	4459	6688	−6723	−6657	−5007	−2499
	16–20	58.67	172	2933	2992	4459	6688	−3962	−3837	−721	4017
	21–25	25.67	382	1283	1309	1951	2926	−3061	−2910	848	6561
	26–30	47.67	213	2383	2431	3623	5434	−891	−693	4258	11,781
	31–35	47.67	124	2383	2431	3623	5434	1368	1613	7755	17,090
	36–40	44.00	97	2200	2244	3344	5016	3470	3760	11,002	22,009
	41–45	40.33	65	2017	2057	3065	4598	5422	5752	14,002	26,542
	46–50	29.33	90	1467	1496	2229	3344	6799	7158	16,142	29,796
Maturity period	51–55	22.00	99	1100	1122	1672	2508	7800	8182	17,715	32,206
	56–60	22.00	99	1100	1122	1672	2508	8802	9205	19,288	34,615
	61–65	18.33	99	917	935	1393	2090	9620	10,042	20,583	36,607
	66–70	14.67	99	733	748	1115	1672	10,255	10,691	21,599	38,180
	71–75	3.67	99	183	187	279	418	10,339	10,779	21,779	38,499
	76–80	14.67	99	733	748	1115	1672	10,974	11,429	22,795	40,073
	81–85	14.67	99	733	748	1115	1672	11,609	12,078	23,811	41,646
	86–90	3.67	99	183	187	279	418	11,693	12,166	23,991	41,965
	91–95	3.67	99	183	187	279	418	11,778	12,255	24,171	42,285

[1] Numbers marked with '- 'indicate deficits. [2] Carbon dioxide absorption during the period (t-CO_2 ha^{-1}). [3] Management costs of implementing the retention approach (including planting, cutting underbrush, tinning, and maintenance). [4] The case of the lowest price of renewable energy credits. [5] The case of the most recent price of renewable energy credits. [6] The case of the average price of renewable energy credits. [7] The case of the highest price of renewable energy credits.

4. Discussion

In this study, we calculated the amount of carbon dioxide absorbed per hectare of Japanese cedar, a common production forests species in Japan. Utilizing the results of this calculation, we examined the possibility of appropriate management of production forests using the carbon credit method, which was introduced by the Tokyo Metropolitan Government. As a result, even with the lowest published carbon credit price of USD 50, the accumulated income and expenditure will be positive 30 years after planting. Beyond that point, the balance of carbon credits and management costs will remain positive, so it can be expected that proper management of production forests will provide a sustainable carbon dioxide absorption function. In Japan, felling occurs 50 years after planting. At that point, the balance is positive, so it is possible to consider a carbon credit system that requires the introduction of a retention approach.

Carbon credits do not subsidize the use of timber or forest management, nor do they build new biomass power plants; they merely change the flow of funds as a system. Therefore, there is almost no new production of carbon dioxide, and this approach is also applicable in countries other than Japan. Appropriate land use and adequate management are some of the infrastructure systems for global environmental conservation [54].

Proper forest management also leads to forest monitoring by which managers can quickly notice small changes in forest conditions. There are many functions that can be activated in a properly managed production forest, such as the establishment of habitats for animals and insects, the preservation of earth water by the soil, and the creation of beautiful landscapes. In degraded forests, proper thinning is not performed, and trees cannot root firmly into the ground, which may cause further damage if disasters happen. In fact, in Japan, the abandonment of production-forests management led to the devastation of forests and increased the likelihood of damage from disasters. In Japan, in 2019, damage from fallen trees during a major typhoon caused power outages for up to 934,900 homes and up to two weeks in areas that included urban areas [55]. In addition, owing to insufficient funds for production-forests management, even when thinning is performed, the thinned wood may remain in the forest, which could also trigger a disaster. Proper production-forests management is important not only for environmental protection but also from a natural risk management perspective.

The use of carbon credits for the carbon dioxide absorption function of planted forest management not only reduces carbon dioxide emissions, but also reduces disaster risk, maintains a beautiful landscape, and provides economic and recreational opportunities. In addition, proper production-forests management ensures the production of quality wood, in case wood demand increases in the future. Thus, this paper shows that carbon credits can be applied to the management of Japanese production forests.

However, even if carbon credits are paid for by the production-forests management, it is necessary to monitor whether appropriate management is performed. Communities located near forests are likely to be suitable for this auditing. Without the construction of these appropriate systems, the carbon credit system may not function effectively.

5. Conclusions

We investigated the possibility of sustainable management and conversion to environmental forests by introducing carbon credits for production forests where the framework of forest management through the use of timber has reached its limit. We found that the management cost of the production forests is compensated by the balance of carbon credits within a maximum of 30 years since planting. The introduction of carbon credits does not introduce new subsidies, nor does it build new facilities such as biomass power plants; it only changes the mechanism. Construction of a biomass power plant is costly, and the transportation of timber also emits carbon dioxide [56,57]. Conversely, the carbon credit method is not only a burden on forest owners and small producers, but there is also almost no burden on the new environment. Moreover, the payment of carbon credits for the carbon-absorption function of forests is limited, but it already exists as a mechanism at this time. By developing the current system, it can be introduced nationwide.

As a policy implication, there are two major advantages of introducing carbon credits to production forests in areas where timber demand is declining sharply: one advantage is that sustainable management can be achieved without depending on subsidies. Mitigating global warming is an urgent task, and the carbon credit system is an indispensable mechanism for reducing greenhouse gases. Utilizing the carbon credit system available in the market, rather than government subsidies, can pave the way for international forest conservation. If independent forest management becomes possible, it will be possible to improve the biased forest structure. The second advantage is emphasizing the value of forests. Production forests were valuable in the past because the trees were used as timber; however, when timber is not used, it becomes difficult to evaluate the value of the forest, and as a result, it tends to lead to abandonment and poor management. Abandonment is likely to induce disasters and lead to the deterioration of the forest function. Millar and Stephenson [58] point out that proper management of forests in temperate climate regions such as Japan can help respond to global warming and minimize the loss of ecosystem services. By introducing carbon credits, the value of forests can be emphasized, thereby motivating forest management. In Japan and other Asian countries where the risk of torrential rains and typhoons is increasing due to climate change, this method can be evaluated as being more effective because the degradation of production forests is likely to amplify natural disaster risks.

Porter-Bolland et al. [59] found that community-managed forests presented lower and less variable annual deforestation rates than protected forests. Based on these results, we believe that carbon credit payments for forest management could be directed not only to landowners but also to local communities. Agrawal and Gibson [60] also noted that effective institutionalization of community-based forest conservation activities requires the availability of sufficient funds for implementing the rules created by local groups. According to our estimation, the introduction of carbon credits resulted in a positive balance when there was sufficient time before logging, making it possible for local communities to evaluate a variety of possibilities, such as the introduction of retention approaches. Scheffer et al. [61] point out that maintaining resilience is important in responding to ecosystem changes caused by global warming and urbanization. From that point of view, the cooperation between the retention approach and carbon credit is important.

Finally, the limitations of this study are as follows: the first is the limitation of data, which is essential for such a study. The exact cost of applying the retention approach to Japanese production forests was unknown. We also lacked the data necessary to make accurate comparisons between production forests and environmental forests. We hope that our research will contribute to the development of a retention approach in Japan in the future. Second, it is possible that the production forests will change to an intensive production forest again; however, it is possible to adopt a retention approach for forests, even if the timber is produced more intensely than before. Third, if the carbon absorption supply increases, the unit price will decrease, and the hypothesis may not hold. However, global warming is advancing, and the demand for carbon credits is expected to increase.

Author Contributions: Conceptualization, N.A.; methodology, N.A. and Y.O.; investigation, N.A.; resources, N.A.; writing, N.A.; review and editing, N.A. and Y.O.; supervision, N.A. and Y.O.; funding acquisition, N.A. All authors have read and agreed to the published version of the manuscript.

Funding: Part of this research was supported by JSPS KAKENHI Grant Number JP 19H02308.

Data Availability Statement: The data used to support the findings of this study are available from the authors upon request.

Conflicts of Interest: The authors declare no conflict of interest.

References

1. Beer, C.; Reichstein, M.; Tomelleri, E.; Ciais, P.; Jung, M.; Carvalhais, N.; Rödenbeck, C.; Arain, M.A.; Baldocchi, D.; Bonan, G.B.; et al. Terrestrial gross carbon dioxide uptake: Global distribution and covariation with climate. *Science* **2010**, *329*, 834–838. [CrossRef]
2. Seto, K.C.; Güneralp, B.; Hutyra, L.R. Global forecasts of urban expansion to 2030 and direct impacts on biodiversity and carbon pools. *Proc. Natl. Acad. Sci. USA* **2012**, *109*, 16083–16088. [CrossRef] [PubMed]
3. Baccini, A.; Goetz, S.J.; Walker, W.S.; Laporte, N.T.; Sun, M.; Sulla-Menashe, D.; Hackler, J.; Beck, P.S.A.; Dubayah, R.; Friedl, M.A.; et al. Estimated carbon dioxide emissions from tropical deforestation improved by carbon-density maps. *Nat. Clim. Chang.* **2012**, *2*, 182–185. [CrossRef]
4. Saatchi, S.S.; Harris, N.L.; Brown, S.; Lefsky, M.; Mitchard, E.T.A.; Salas, W.; Zutta, B.R.; Buermann, W.; Lewis, S.L.; Hagen, S.; et al. Benchmark map of forest carbon stocks in tropical regions across three continents. *Proc. Natl. Acad. Sci. USA* **2011**, *108*, 9899–9904. [CrossRef] [PubMed]
5. Lindenmayer, D.B.; Franklin, J.F.; Lõhmus, A.; Baker, S.C.; Bauhus, J.; Beese, W.; Brodie, A.; Kiehl, B.; Kouki, J.; Pastur, G.M.; et al. A major shift to the retention approach for forestry can help resolve some global forest sustainability issues. *Conserv. Lett.* **2012**, *5*, 421–431. [CrossRef]
6. Gustafsson, L.; Baker, S.C.; Bauhus, J.; Beese, W.J.; Brodie, A.; Kouki, J.; Lindenmayer, D.B.; Lohmus, A.; Pastur, G.M.; Messier, C.; et al. Retention Forestry to Maintain Multifunctional Forests: A World Perspective. *Bioscience* **2012**, *62*, 633–645. [CrossRef]
7. Hartley, M.J. Rationale and methods for conserving biodiversity in plantation forests. *For. Ecol. Manag.* **2002**, *155*, 81–95. [CrossRef]
8. Mori, A.S.; Kitagawa, R. Retention forestry as a major paradigm for safeguarding forest biodiversity in productive landscapes: A global meta-analysis. *Biol. Conserv.* **2014**, *175*, 65–73. [CrossRef]
9. Pastur, G.M.; Lencinas, M.V.; Cellini, J.M.; Peri, P.L.; Soler Esteban, R.S. Timber management with variable retention in Nothofagus pumilio forests of Southern Patagonia. *For. Ecol. Manag.* **2009**, *258*, 436–443. [CrossRef]
10. Ozaki, K.; Akashi, N.; Unno, A.; Sato, S.; Sayama, K.; Nagasaka, A.; Nagasaka, Y.; Yamada, K.; Yamaura, Y. Retention forestry to manage forests for timber production and biodiversity conservation—A review to promote the effective application in Japan. *Jpn. J. Ecol.* **2018**, *68*, 101–123.
11. Barreto, P.; Amaral, P.; Vidal, E.; Uhl, C. Costs and benefits of forest management for timber production in eastern Amazonia. *Forest Ecol. Manag.* **1998**, *108*, 9–26. [CrossRef]
12. Arnott, J.T.; Beese, W.J. Alternatives to clearcutting in BC Coastal Montane Forests. *For. Chron.* **1997**, *73*, 670–678. [CrossRef]
13. Ministry of Environment, Japan. Offset Credit (J-VER) System. 2020. Available online: http://offset.env.go.jp/j-ver/pref.html (accessed on 15 March 2021). (In Japanese)
14. Hurteau, M.D.; Koch, G.W.; Hungate, B.A. Carbon protection and fire risk reduction: Toward a full accounting of forest carbon offsets. *Front. Ecol. Environ.* **2008**, *6*, 493–498. [CrossRef]
15. Malmsheimer, R.W.; Bowyer, J.L.; Fried, J.S.; Gee, E.; Izlar, R.L.; Miner, R.A.; Munn, I.A.; Oneil, E.; Stewart, W.C. Managing forests because carbon matters: Integrating energy, products, and land management policy. *J. For.* **2011**, *109*, 7–48.
16. Bradshaw, C.J.A.; Bowman, D.M.J.S.; Bond, N.R.; Murphy, B.P.; Moore, A.D.; Fordham, D.A.; Thackway, R.; Lawes, M.J.; McCallum, H.; Gregory, S.D.; et al. Brave new green world—Consequences of a carbon economy for the conservation of Australian biodiversity. *Biol. Conserv.* **2013**, *161*, 71–90. [CrossRef]
17. Fletcher, L.S.; Kittredge, D.; Stevens, T. Forest Landowners' Willingness to Sell Carbon Credits: A Pilot Study. *N. J. Appl. For.* **2009**, *26*, 35–37. [CrossRef]
18. Preece, N.D.; van Oosterzee, P.V.; Hidrobo Unda, G.C.; Lawes, M.J. National carbon model not sensitive to species, families and site characteristics in a young tropical reforestation project. *For. Ecol. Manag.* **2017**, *392*, 115–124. [CrossRef]
19. Intergovernmental Panel on Climate Change (IPCC). IPCC Special Report on Land Use Land-Use Change and Forestry. 1998. Available online: https://www.grida.no/climate/ipcc/land_use/index.htm (accessed on 15 March 2021).
20. Food and Agriculture Organization of the United Nations (FAO). *Global Forest Resources Assessment 2020 Main Report*; FAO: Rome, Italy, 2020.
21. Butchart, S.H.M.; Walpole, M.; Collen, B.; Van Strien, A.; Scharlemann, J.P.W.; Almond, R.E.A.; Baillie, J.E.M.; Bomhard, B.; Brown, C.; Bruno, J.; et al. Global biodiversity: Indicators of recent declines. *Science* **2010**, *328*, 1164–1168. [CrossRef] [PubMed]
22. Lindner, M.; Maroschek, M.; Netherer, S.; Kremer, A.; Barbati, A.; Garcia-Gonzalo, J.; Seidl, R.; Delzon, S.; Corona, P.; Kolström, M.; et al. Climate change impacts, adaptive capacity, and vulnerability of European forest ecosystems. *For. Ecol. Manag.* **2009**, *259*, 698–709. [CrossRef]
23. Seidl, R.; Thom, D.; Kautz, M.; Martin-Benito, D.; Peltoniemi, M.; Vacchiano, G.; Wild, J.; Ascoli, D.; Petr, M.; Honkaniemi, J.; et al. Forest disturbances under climate change. *Nat. Clim. Chang.* **2017**, *7*, 395–402. [CrossRef] [PubMed]
24. United Nations. Transforming Our World: The 2030 Agenda for Sustainable Development. 2015. Available online: https://sustainabledevelopment.un.org/content/documents/21252030%20Agenda%20for%20Sustainable%20Development%20web.pdf (accessed on 15 March 2021).
25. Chazdon, R.L. Beyond deforestation: Restoring forests and ecosystem services on degraded lands. *Science* **2008**, *320*, 1458–1460. [CrossRef]

26. Groot, A.; Lussier, J.-M.; Mitchell, A.K.; MacIsaac, D.A. A silvicultural systems perspective on changing Canadian forestry practices. The Forestry Chronicle. *For. Chron.* **2005**, *81*, 50–55. [CrossRef]
27. Asner, G.P.; Powell, G.V.N.; Mascaro, J.; Knapp, D.E.; Clark, J.K.; Jacobson, J.; Kennedy-Bowdoin, T.; Balaji, A.; Paez-Acosta, G.; Victoria, E.; et al. High-resolution forest carbon stocks and emissions in the Amazon. *Proc. Natl. Am. Sci. USA* **2010**, *107*, 16738–16742. [CrossRef] [PubMed]
28. Harris, N.L.; Brown, S.; Hagen, S.C.; Saatchi, S.S.; Petrova, S.; Salas, W.; Hansen, M.C.; Potapov, P.V.; Lotsch, A. Baseline Map of Carbon Emissions from Deforestation in Tropical Regions. *Science* **2012**, *336*, 1573–1576. [CrossRef]
29. Lewis, S.L.; Lopez-Gonzalez, G.; Sonke, B.; Affum-Baffoe, K.; Baker, T.R.; Ojo, L.O.; Phillips, O.L.; Reitsma, J.M.; White, L.; Comiskey, J.A.; et al. Increasing carbon storage in intact African tropical forests. *Nature* **2009**, *457*, 1003–1007. [CrossRef]
30. Keith, H.; Mackey, B.G.; Lindenmayer, D.B. Re-evaluation of forest biomass carbon stocks and lessons from the world's most carbon-dense forests. *PNAS* **2009**, *106*, 11635–11640. [CrossRef] [PubMed]
31. Matsumoto, M. Carbon accumulation and carbon absorption by Japanese forests. *Jpn. For. Soc.* **2001**, *33*, 30–36. (In Japanese) [CrossRef]
32. Chave, J.; Andalo, C.; Brown, S.; Cairns, M.A.; Chambers, J.Q.; Eamus, D.; Folster, H.; Fromard, F.; Higuchi, N.; Kira, T.; et al. Tree allometry and improved estimation of carbon stocks and balance in tropical forests. *Oecologia* **2005**, *145*, 87–99. [CrossRef]
33. Jenkins, J.C.; Chojnacky, D.C.; Heath, L.S.; Birdsey, R.A. National-scale biomass estimators for United States tree species. *For. Sci.* **2003**, *49*, 12–35.
34. Forestry and Forest Products Research Institute (FFPRI). How to Capture Forest Carbon Absorption. Available online: http://www.ffpri.affrc.go.jp/research/dept/22climate/kyuushuuryou/ (accessed on 15 March 2021). (In Japanese)
35. Canadell, J.G.; Raupach, M.R. Managing forests for climate change mitigation. *Science* **2008**, *320*, 1456–1457. [CrossRef]
36. Fahey, T.J.; Woodbury, P.B.; Battles, J.J.; Goodale, C.L.; Hamburg, S.P.; Ollinger, S.V.; Woodall, C.W. Forest carbon storage: Ecology, management, and policy. *Front. Ecol. Environ.* **2010**, *8*, 245–252. [CrossRef]
37. Kurz, W.A.; Dymond, C.C.; White, T.M.; Stinson, G.; Shaw, C.H.; Rampley, G.J.; Smyth, C.; Simpson, B.N.; Neilson, E.T.; Trofymow, J.A.; et al. CBM-CFS3: A model of carbon-dynamics in forestry and land-use change implementing IPCC standards. *Ecol. Model.* **2009**, *220*, 480–504. [CrossRef]
38. Morse, W.C.; Schedlbauer, J.L.; Sesnie, S.E.; Finegan, B.; Harvey, C.A.; Hollenhorst, S.J.; Kavanagh, K.L.; Stoian, D.; Wulfhorst, J.D. Consequences of environmental service payments for forest retention and recruitment in a Costa Rican biological corridor. *Ecol. Soc.* **2009**, *14*, 23. [CrossRef]
39. Kayo, C.; Amano, K.; Shimada, K. Evaluation of the forest management procedure in Japan based on long-term carbon mass balance. *Environ. Syst. Res.* **2006**, *34*, 235–243. [CrossRef]
40. Ellison, D.; Lundblad, M.; Petersson, H. Carbon accounting and the climate politics of forestry. Environ. *Sci. Policy* **2011**, *14*, 1062–1078. [CrossRef]
41. Ajiki, K. Changes of National Forest Management after the 1980's and Regional Characteristic. Bulletin Faculty of Humanities and Social Sciences. *Mie Univ.* **2010**, *27*, 1–21. Available online: https://mie-u.repo.nii.ac.jp/?action=repository_uri&item_id=1539&file_id=17&file_no=1 (accessed on 15 March 2021). (In Japanese).
42. Ministry of Agriculture, Forestry and Fisheries, Japan. Forestry Management Statistics Survey, Confirmation Report 2013 Forestry Management Statistics Survey Report. Available online: https://www.e-stat.go.jp/stat-search/file-download?statInfId=000029151864&fileKind=0 (accessed on 15 March 2021). (In Japanese)
43. Ministry of Foreign Affairs, Japan. Submission of Japan's Intended Nationally Determined Contribution. Available online: https://www.mofa.go.jp/mofaj/files/000090898.pdf (accessed on 15 March 2021).
44. Tokyo Metropolitan Government. Guidelines for Calculating Renewable Energy Credits in Total Volume Reduction Obligations and Emissions Trading Schemes. Available online: https://www.kankyo.metro.tokyo.lg.jp/climate/large_scale/rules/.cat9740.files/201904_guideline_saiene_credit_santei.pdf (accessed on 15 March 2021). (In Japanese)
45. Tokyo Metropolitan Government. Policy from 2020 on Major Systems for Climate Change Countermeasures. Available online: https://www.kankyo.metro.tokyo.lg.jp/climate/large_scale/overview/after2020/outline.files/bessi02_.pdf (accessed on 15 March 2021). (In Japanese)
46. Tokyo Metropolitan Government. Tokyo Emissions Trading Seminar: About the Total Amount Reduction Obligation and the Transaction Price of the Emissions Trading System. 2020. Available online: https://www.kankyo.metro.tokyo.lg.jp/climate/large_scale/trade/index.files/sateikakaku1216.pdf (accessed on 15 March 2021). (In Japanese)
47. United Nations University Institute for the Advanced Study of Sustainability. The International Partnership for the Satoyama initiative: Information Booklet and 2016 Annual Report. Available online: https://satoyama-initiative.org/wp-content/uploads/2018/02/IPSI-2016-AR-and-booklet-final-compressed.pdf (accessed on 15 March 2021).
48. Takeuchi, K. Rebuilding the relationship between people and nature: The Satoyama Initiative. *Ecol. Res.* **2010**, *25*, 891–897. [CrossRef]
49. Forestry Agency, Management Division, Forestry Labor Countermeasures Office, Japan. Status of Securing Forestry Labor Force. Available online: https://www.mhlw.go.jp/file/05-Shingikai-11201000-Roudoukijunkyoku-Soumuka/0000185117.pdf (accessed on 15 March 2021). (In Japanese)
50. Ministry of Land, Infrastructure, Transport and Tourism, Japan. On the Situation of Living in Japan. pp. 19–20. Available online: https://www.mlit.go.jp/policy/shingikai/content/001308853.pdf (accessed on 15 March 2020). (In Japanese)

51. Ministry of Agriculture, Forestry and Fisheries, Japan. Forest and Forestry White Paper: Trends in Forests and Forestry in 2018. 2018; p. 97. Available online: https://www.rinya.maff.go.jp/j/kikaku/hakusyo/30hakusyo/attach/pdf/zenbun-13.pdf (accessed on 15 March 2021). (In Japanese)
52. Ministry of Agriculture, Forestry and Fisheries, Japan. Forest Area by Tree Species and Age Class in 2017. Available online: https://www.rinya.maff.go.jp/j/keikaku/genkyou/h29/attach/pdf/4-22.pdf (accessed on 15 March 2021). (In Japanese)
53. Ministry of Agriculture, Forestry and Fisheries, Japan. Forest Rate and Planted Forest Rate by Prefecture (as of 31 March 2017). Available online: https://www.rinya.maff.go.jp/j/kikaku/hakusyo/29hakusyo/attach/pdf/zenbun-4.pdf (accessed on 15 March 2021). (In Japanese)
54. Foley, J.A.; De Fries, R.; Asner, G.P.; Barford, C.; Bonan, G.; Carpenter, S.R.; Chapin, F.S.; Coe, M.T.; Daily, G.C.; Gibbs, H.K.; et al. Global consequences of land use. *Science* **2005**, *309*, 570–574. [CrossRef] [PubMed]
55. Tokyo Electric Power Company Holdings. Review of the Response to Power Outage Restoration Following Typhoon No. 15 (Interim Report). 2019. Available online: https://www.tepco.co.jp/press/release/2019/pdf4/191031j0201.pdf (accessed on 15 March 2021). (In Japanese)
56. Guo, M.; Song, W.; Buhain, J. Bioenergy and biofuels: History, status, and perspective. *Renew. Sustain. Energy Rev.* **2015**, *42*, 712–725. [CrossRef]
57. Schulze, E.-D.; Körner, C.; Law, B.E.; Haberl, H.; Luyssaert, S. Large-scale bioenergy from additional harvest of forest biomass is neither sustainable nor greenhouse gas neutral. *GCB Bioenergy* **2012**, *4*, 611–616. [CrossRef]
58. Millar, C.I.; Stephenson, N.L. Temperate forest health in an era of emerging mega disturbance. *Science* **2015**, *349*, 823–826. [CrossRef]
59. Porter-Bolland, L.; Ellis, E.A.; Guariguata, M.R.; Ruiz-Mallén, I.; Negrete-Yankelevich, S.; Reyes-García, V. Community managed forests and forest protected areas: An assessment of their conservation effectiveness across the tropics. *For. Ecol. Manag.* **2012**, *268*, 6–17. [CrossRef]
60. Agrawal, A.; Gibson, C.C. Enchantment and disenchantment: The role of community in natural resource conservation. *World Dev.* **1999**, *27*, 629–649. [CrossRef]
61. Scheffer, M.; Carpenter, S.; Foley, J.A.; Folke, C.; Walker, B. Catastrophic shifts in ecosystems. *Nature* **2001**, *413*, 591–596. [CrossRef] [PubMed]

forests

MDPI

Article

The Temporal and Spatial Evolution of Ecosystem Service Synergy/Trade-Offs Based on Ecological Units

Teng Niu [1], Jiaxin Yu [2], Depeng Yue [1,*], Linzhe Yang [1], Xueqing Mao [1], Yahui Hu [1] and Qianqian Long [1]

[1] Beijing Key Laboratory for Precision Forestry, Beijing Forestry University, Beijing 100083, China; niuteng21@bjfu.edu.cn (T.N.); ylz15235456201@163.com (L.Y.); maoxq18@bjfu.edu.cn (X.M.); hu_yahui_one@163.com (Y.H.); 15983139552@163.com (Q.L.)

[2] Shijiazhuang Forestry Bureau Office, Shijiazhuang 050051, China; lyjbgs208@126.com

* Correspondence: yuedepeng@126.com; Tel.: +86-010-62337585

Abstract: "Two ecological barriers and three shelters" (TEBTS), which has the effect of relieving ecological pressure, is the national ecological security pattern in China. Calculating the value of TEBTS ecosystem services, clarifying the synergy/trade-off relationships between ecosystem services, and maximizing the value of regional ecosystem services are of great significance for maintaining the security of the ecological civilization. At present, the research on ecosystem service synergy/trade-off has become the frontier field of ecology and related disciplines at home and abroad, and many research results have been obtained. However, there is still room and significance for continuing research to think about the synergy/trade-off relationship of ecosystems from the perspective of temporal and spatial heterogeneity: clarifying the spatial scope and spatial transmission characteristics of ecosystem service synergy/trade-off; exploring the trend of ecosystem service synergy/trade-off, and simulating the dynamic characteristics of natural factors affecting ecosystem services; and analyzing the characteristics of different spatial attributes that lead to the synergy/trade-off of ecosystem services. In this study, the Songhua River Basin (SRB), where the NFB is located, is used as the research area, the ecosystem services are simulated through the ecosystem assessment model, ecological unit (EU) is constructed as a research carrier, which is used to define the spatial scope of ecosystem services, and the influence of spatial characteristics and attribute characteristics on the change trend of the ecosystem service synergy/trade-off relationship is analyzed. The research found that water retention, soil conservation, and biodiversity did not change much from 2000 to 2015, and these ecosystem services have a greater value in the NFZ. The amount of carbon sequestration increased rapidly from 2010 to 2015. Crop production showed an increasing trend year by year. As the main grain production area, the Songnen Plain provides the main crop production function, which is greatly affected by humans. In the spatial characteristic, water retention, soil sequestration, and biodiversity present a very significant synergistic relationship, which is manifested in the obvious high-value aggregation characteristics in the NFZ, and crop production and the other four types of ecosystem services are in a trade-off relationship. At the time scale, the four types of ecosystem services, including water retention, soil conservation, biodiversity, and carbon sequestration, are synergistic, and crop production and water retention are synergistic. The vegetation types exhibiting a synergy/trade-off relationship are mainly broad-leaved forests, and the soil types are mainly luvisols and phaeozems. These EUs are mainly distributed in the NFZ and have spatial topological characteristics: the area and circumference of these EUs are smaller, the radius of gyration is also significantly smaller than that of other EUs, and the shape is more regular. By focusing on the spatial aggregation characteristics and changing trends of the ecosystem service synergy/trade-off and clarifying the influencing factors of the ecosystem service synergy/trade-off, the ecosystem services can be integrated, and the ecosystem can be optimized. Thus, the value of regional ecosystem services can be maximized, and a certain data foundation and theoretical support can be provided for major projects, such as ecological restoration and ecological environment governance, which is of great significance for improving the pattern of ecological security.

Keywords: ecosystem services; ecological unit; synergy/trade-off; spatial structure; attribute characteristics

Citation: Niu, T.; Yu, J.; Yue, D.; Yang, L.; Mao, X.; Hu, Y.; Long, Q. The Temporal and Spatial Evolution of Ecosystem Service Synergy/Trade-Offs Based on Ecological Units. *Forests* **2021**, *12*, 992. https://doi.org/10.3390/f12080992

Academic Editors: Juan F. Fernández-Manjarrés and Roxane Sansilvestri

Received: 7 June 2021
Accepted: 20 July 2021
Published: 26 July 2021

1. Introduction

With the increase in population and rapid development of the social economy, the current ecological environment in China is becoming more and more severe. The pressure of human society on resources and the environment is increasing, which increasingly threatens the ecological security of the country and its regions [1]. In order to relieve the ecological pressure, a national barrier plan was proposed based on the main functional zoning [2]. A national ecological security strategic pattern, with "Two ecological barriers and three shelters" (TEBTS) as the mainstay, came into being [3]. The TEBTS ecological security strategic pattern is an important part of the "three strategic patterns" and an important guarantee pattern of the urbanization pattern strategy and the agricultural strategic pattern. The TEBTS ecological security strategic pattern is to build TEBTS and important river systems as the framework, with other national key ecological function areas as important support, and national development prohibited areas as an important component of the ecological security strategic pattern. TEBTS is an important part of building a national ecological security strategic pattern in the construction of ecological civilization, and a national ecological security pattern has been established. Providing ecosystem services is an important manifestation of the ecological security pattern. TEBTS is key to regional ecological security, and the ecosystem plays an important role in the region [4]. Therefore, a well-structured ecosystem is obviously the main body and the first element of an ecological barrier [5–8]. Each barrier zone of the TEBTS has a principal ecosystem service. Due to the different needs of human beings, the ecosystem services provided by the barrier zone have a variety of extended features. In recent years, the research on the calculation of ecosystem services has tended toward diversification, and its quantitative evaluation methods are summarized into three categories: the physical measurement method, the energy method, and the value measurement method. Specific models for calculating specific ecosystem services include the water balance equation, general soil loss equation, WEPP model, SWAT model, InVEST model, ARIES model, etc. Among them, the trade-off and synergy between multiple ecosystem services has always been an important topic for mainstream scholars. Due to the diversity of ecosystem service types, the imbalance of spatial distribution, and the selectivity of human use, the relationships between ecosystem services have undergone dynamic changes, which are manifested in the form of trade-offs and synergy between mutual gains. Ecosystem services have intertwined and complex nonlinear relationships which are superimposed on the different preferences of humans concerning the choice of ecosystem services [9]. Changes in ecosystem pattern-process-function services lead to a synergy/trade-off between ecosystem services [10–12]. In practical applications, methods such as mathematical statistics, spatial mapping, scenario simulation, multiobjective decision making, and service liquidity analysis are generally used to carry out research on the synergy/trade-off between different service types at the time and space scales.

In recent years, conducting the synergy/trade-off analysis of ecosystem services and understanding the relationships between ecosystem services has become an important basis for ecological management decision making and regulation, and it is applied to agriculture, fishery production [13,14], forest management, marine space planning, energy management, etc. [15–18]. However, the synergy/trade-off of ecosystem services has a relatively complex temporal and spatial scale, and the difficulty lies in how to characterize the interaction between the structure, process, function, and service changes of ecosystems at different scales and their influencing factors. Scholars have also carried out a lot of research on this. Chan et al. found through GIS spatial analysis that there is only a weak correlation between the priority areas for biodiversity conservation and the six ecosystem service supply areas in the California Central Coast Ecological Zone [19]. Egoh et al. used graphs to characterize the supply of five ecosystem services in South Africa (surface water supply, water flow regulation, soil accumulation, soil retention, and carbon storage) and then assessed the relationship between them [20]. Raudsepp Hearne et al. conducted the spatial mapping and cluster analysis of 12 types of ecosystem services, identified 6 types of ecosystem service clusters, and finally identified the types and regions of synergy and trade-offs between different services [21]. Based on the production possibility boundary method, Chen et al. conducted a quantitative analysis on the relationship between ecosystem services in the Weihe River Basin. The study showed that there is a trade-off relationship between water retention and carbon sequestration and between water retention and biodiversity, and there is synergy between carbon sequestration and biodiversity [22]. Through partial correlation analysis, Sun et al. explored the trade-off and synergy between the ecosystem services in Yan'an. They found that there is a trade-off relationship between soil conservation and crop production and between water retention and NPP, and there is a synergistic relationship between NPP and soil conservation and between NPP and water retention [23].

However, most scholars have not conducted outstanding research on the spatial heterogeneity of ecosystem service synergy/trade-offs. From a spatial perspective, the response to changes in ecosystem services is based on irregular geographic units. Clarifying the spatial boundaries and spatial characteristics of ecosystem services is an important research direction for the integration of ecosystem services and geography. Geographical units are patches with spatial characteristics based on landscape types.

The ecosystem service evaluation models currently applied are all based on grid units in space, which cannot fully express the spatial heterogeneity of terrain, vegetation, and soil. In a complex geographic environment, it cannot integrate topography and other factors into the study of ecosystem services. The synergy/trade-off analysis of ecosystem services should not be based on a regular grid unit as a carrier, but should be a geographical unit with independent characteristics. Regardless of their size or shape, it should be considered whether the geographic features in the geographic unit are consistent. The grids with consistent geographic features are summarized as a geographic unit, and each geographic unit has a unique spatial structure and topological relationship. Only by using this geographic unit for analysis can we better evaluate the ecosystem service synergy/trade-off relationship as a whole. Therefore, it is urgent to establish a method based on the onsite geomorphology to evaluate the synergy/trade-off relationship of ecosystem services. An ecological unit (EU) is a spatial patch delineated based on factors such as hydrology, land cover, vegetation types, and soil types. EUs can reflect the spatial changes of ecosystem service elements and can describe the impact of changes in underlying surface features on ecosystem services [24,25]. EUs are land surface complexes with the same terrain distribution, vegetation types, and soil conditions. They are also geographic units that integrate the spatial heterogeneity of elements such as terrain, soil, and land use patterns. EUs include a unique water outlet, which can effectively reflect the law of water movement in the basin. In EUs, the movement of water and sediment will be directed to the only water outlet. The water retention function and the soil conservation function can be reflected by EUs. Meanwhile, since the ecological unit is constructed with unique topography, vegetation, soil, and land features, ecosystem services, such as carbon sequestration and biodiversity, are also unique. At the spatial scale, EUs are used as carriers for studying the synergy/trade-off relationship of ecosystem services and express the aggregation characteristics of ecosystem services. At the time scale, the influence of the spatial characteristics of EUs on the synergy/trade-off of ecosystem services needs to be further studied, and the theory of landscape pattern needs to be applied to the analysis of the structure of EUs. The landscape pattern is a feature of the spatial pattern, that is, the spatial arrangement and combination of landscape elements of different sizes and shapes. It includes the type, number, and spatial distribution and configuration of the constituent units, which are a concrete manifestation of spatial heterogeneity. The pattern index constructed using the theory of landscape pattern can effectively reflect the spatial information of EUs. EUs can evaluate the characteristics of the spatial structure, combining the attributes of the EUs themselves and discussing the influence of the characteristics of the EU on the ecosystem service synergy/trade-off relationship in a certain period of time.

Therefore, the use of ecological units in the study of ecosystem service synergy/tradeoffs has theoretical and practical foundations.

This study takes the SRB as the research area, calculating the value of ecosystem services in 2000, 2005, 2010, and 2015, respectively, constructing EUs and using EUs as carriers to conduct synergy/trade-off spatiotemporal analysis, and at the spatial scale, using bivariate local autocorrelation to determine the ecosystem service synergy/trade-off relationship and clarify the clustering characteristics of ecosystem services. Through correlation analysis, the study distinguishes the synergy/trade-off relationship of ecological service functions at the time scale with significant characteristics. At the same time, it explores the relationship between the attribute characteristics and spatial characteristics in the ecological unit and the synergy/trade-off of ecosystem services.

2. Study Area

The SRB (Figure 1) covers an area of 546,000 km^2, which belongs to the three provinces and autonomous regions of Inner Mongolia, Jilin, and Heilongjiang, of which 61% are mountainous areas, 15% are hills, and 24% are plains. The Sanjiang and Songnen Plains are in the east and west of the river basin, the land is fertile, and the grassland is continuous. The SRB has an obvious continental monsoon climate, with hot and rainy summers and cold and dry winters, and the four seasons have obvious differences [26]. There are obvious spatial differences in the average precipitation in the basin over many years. The rainfall in the mountainous areas in the southeast is between 700 and 900 mm, while the western region is relatively dry, with an annual rainfall of only about 400 mm. The average annual runoff of the SRB over many years reached 76.2 billion m^3. The annual average runoff depth of the basin over many years is 134.3 mm, and the spatial distribution characteristics are basically consistent with the annual precipitation. In the highest mountain area, it can reach about 500 mm, while in the Songnen Plain, it is only 20–30 mm. In the SRB, there are eleven principal types of soils, including luvisols, phaeozems, and chernozems. The vegetation types in the SRB are mainly divided into three categories: forests in mountains and hilly areas and grasslands and crops in the Northeast Plain. Due to the relationship between the latitude and climate, the Great Khingan and Lesser Khingan Mountains have coniferous forests in the sub-frigid zone, and there are cold and temperate coniferous and broad-leaved forests in the Changbai Mountains. In the northwest of the Songnen Plain, forest surrounds the steppe [27,28].

Figure 1. Location of the study area.

3. Data and Methodology

3.1. Data

The boundaries of the SRB and the NFZ are provided by the "National Scale Ecosystem Service Evaluation and Analysis Model System". The Digital Elevation Model (DEM) is obtained from the Geospatial Data Cloud (http://www.gscloud.cn/ accessed on 25 September 2020). The data sources required by the ecosystem service assessment model are shown in Table 1.

3.2. Approaches

This study (Figure 2) calculates water retention based on the theory of water balance, uses the RUSLE model to calculate soil conservation, uses the CASA model to calculate carbon sequestration, calculates biodiversity based on indicators such as the biological abundance index, and calculates the crop production based on the grain yield in the statistical year-book. It calculates ecosystem services in 2000, 2005, 2010, and 2015, respectively. EUs are delineated by terrain, vegetation type, and soil type, and EUs are used as research carriers. At the spatial scale, spatial autocorrelation expresses the ecosystem service synergy/trade-off relationship. The synergy relationship is embodied in ecosystem service high–high aggregation and low–low aggregation, and the trade-off relationship is embodied as high–low aggregation. Ecosystem services at the time scale are calculated by a unit-by-unit analysis method, and the impact of attribute characteristics and spatial topological characteristics on ecosystem service synergy/trade-off is analyzed.

Table 1. The data sources required by the ecosystem service assessment model.

Ecosystem Services	Required Data	Data Sources
Water retention	Precipitation	Obtained by interpolation of national meteorological station data downloaded from China Meteorological Data Network (http://data.cma.cn/, accessed on 15 September 2020)
	Actual evapotranspiration	From Institute of Mountain Hazards and Environment, Chinese Academy of Sciences
	Runoff	From Institute of Mountain Hazards and Environment, Chinese Academy of Sciences
Soil conservation	Slope length	Use ArcGIS software to process DEM data acquisition
	Slope	Use ArcGIS software to process DEM data acquisition
	Normalized Difference Vegetation Index	Obtained by Landsat image band calculation
Carbon sequestration	Average monthly precipitation	From China Meteorological Data Network (http://data.cma.cn/, accessed on 15 September 2020)
	Monthly average temperature	From China Meteorological Data Network (http://data.cma.cn/, accessed on 15 September 2020)
	Sun radiation	From China Meteorological Data Network (http://data.cma.cn/, accessed on 15 September 2020)
	Vegetation distribution	From Institute of Mountain Hazards and Environment, Chinese Academy of Sciences
	Soil distribution	From Institute of Mountain Hazards and Environment, Chinese Academy of Sciences
	Normalized Difference Vegetation Index	Obtained by Landsat image band calculation
Biodiversity	Richness of rare and endangered biological species	From Institute of Mountain Hazards and Environment, Chinese Academy of Sciences
	Biological abundance index	From Institute of Mountain Hazards and Environment, Chinese Academy of Sciences
	Normalized Difference Vegetation Index	Obtained by Landsat image band calculation
Crop production	The per unit area yield of grain	From the National County Statistical Yearbook
	Climate data	From China Meteorological Data Network (http://data.cma.cn/, accessed on 15 September 2020)
	Slope	Use ArcGIS software to process DEM data acquisition

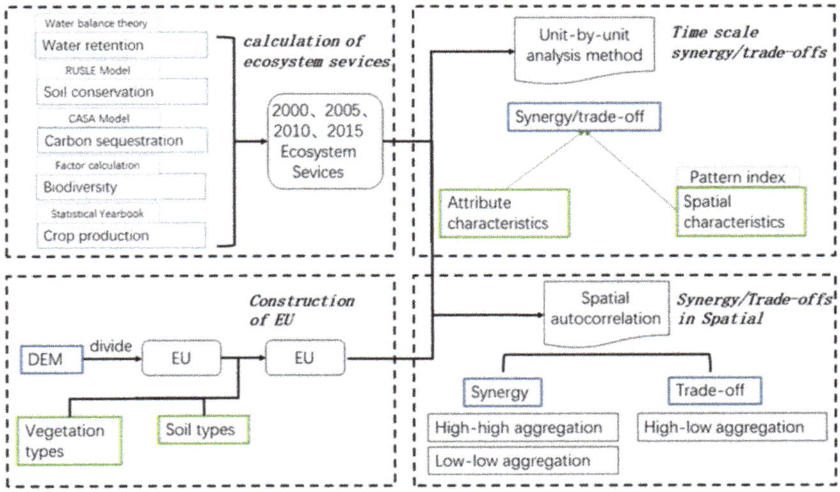

Figure 2. The research framework of the ecosystem services synergy/trade-off.

3.2.1. Ecosystem Service Assessment Method

Water Retention Estimation

The improved water balance theory is used to evaluate the ecosystem service of water retention [29]. The theory believes that water retention is precipitation minus evapotranspiration (including canopy transpiration, canopy evaporation, and soil evaporation) and storm runoff. The whole process begins with atmospheric precipitation, after which part of the precipitation is intercepted by the canopy, and the other part penetrates the canopy and enters the soil [30,31].

The water retention amount (WR) is calculated according to the water balance equation, namely:

$$WR = P - ET - R = P - \left(E_{canopy} + E_{soil} + T + E_{sub}\right) - R \quad (1)$$

In the formula, P is the precipitation, ET is the sum of the transpiration and evaporation (including transpiration T, canopy intercepted evaporation E_{canopy}, soil evaporation E_{soil}, and snow sublimation E_{sub}), and R is the surface runoff.

Soil Conservation Estimation

The soil conservation service function evaluation is based on the RUSLE model [32–34]. It is assumed that the soil conservation amount is the difference between the potential soil erosion amount and the actual soil erosion amount, that is, the difference between the actual soil erosion amount in the maximum soil erosion amount area without vegetation cover. The potential soil erosion is the amount of soil erosion determined by natural factors, such as climate, soil, topography, etc., that is, the state of soil erosion under the assumption that there is no vegetation protection. The actual soil erosion is the amount of soil erosion determined by factors such as climate, soil, topography, vegetation, etc., that is, the state of soil erosion under the existing vegetation coverage [35–37].

The difference between potential soil erosion and actual soil erosion is used to characterize the soil conservation function of the ecosystem. Considering the spatial and temporal scale, model complexity, and data availability, the amount of soil erosion is estimated using the Universal Soil Loss Equation (RUSLE). The model form is as follows:

Actual soil erosion:

$$A_{actual\ erosion} = R \times K \times L \times S \times C \quad (2)$$

Potential soil erosion:

$$A_{potential\ erosion} = R \times K \times L \times S \quad (3)$$

Soil conservation amount:

$$A_{soil\ erosion} = A_{potential\ erosion} - A_{actural\ erosion} = R \times K \times L \times S \times (1 - C) \quad (4)$$

In the formula, $A_{actual\ erosion}$ is the actual amount of soil erosion per unit area, and the unit is t/ha; $A_{potential\ erosion}$ is the potential soil erosion amount per unit area, and the unit is t/ha; $A_{soil\ erosion}$ is the amount of soil conservation per unit area, and the unit is t/ha; R is the rainfall erosivity factor, expressed by the multi-year average annual rainfall erosivity index; K is the soil erodibility factor, expressed as the amount of soil loss per unit area formed by the unit rainfall erosivity in the standard plot; L is the slope length factor (dimensionless); S is the slope factor (dimensionless); and C is the vegetation cover factor (dimensionless).

Carbon Sequestration Estimation

In the carbon sequestration estimation, 1 kg of carbon is considered to be equivalent to 2.2 kg of organic matter. The carbon sequestration of the ecosystem can be obtained from the total amount of vegetation NPP

$$W_{CO_2} = NPP \times 2.2 \quad (5)$$

In the formula, W_{CO_2} represents the fixed amount of CO_2 per unit area of an ecosystem (g/m^2), and NPP represents the annual vegetation NPP per unit area of the ecosystem (g/m^2).

There are many models for NPP estimation [38], and the prototype CASA model was proposed by Monteith. Later, Potter and Field et al. improved Monteith's model. Due to its simple structure, and because the parameters can be obtained through remote sensing, the CASA model is widely used in NPP estimation at global and regional scales.

Biodiversity Estimation

Biodiversity is calculated by a comprehensive calculation of the biological abundance index, vegetation coverage, and number of rare and endangered animals and plants, expressed by the formula:

$$BIO = B \times VC \times TS \quad (6)$$

In the formula, BIO is the biodiversity maintenance function index, dimensionless; B is the biological abundance index, dimensionless, which is used to assign values according to the land use type; VC is the vegetation coverage, %; and TS is the richness of rare and endangered biological species, dimensionless.

Crop Production Estimation

The crop production in the statistical yearbook is used to establish a statistical relationship between geographical environment parameters in order to realize the spatialization of crop production. Statistics show that the annual precipitation, annual average temperature, and topographical undulations of the grain yield levels are significantly correlated, and the confidence level has passed the 0.05 test. Thus, the statistical model is established as follows:

$$SCrop = 0.00047P + 3.70805 \times e^{0.0285Ta} - 0.00075Range - 0.26622 \quad (7)$$

where $SCrop$ is the simulated crop production, t/ha; P is the annual precipitation, mm; Ta is the average annual temperature (°C); and $Range$ is the topographic undulation (m).

3.2.2. Construction of Ecological Units

The results of the calculation of the ecosystem services are presented in the form of raster data. However, in reality, due to multiple factors, such as spatial differences and irregular land use types, the division of the ecosystem services must not be of a regular shape, like a grid. Therefore, we need to find a brand-new unit division method, transform the data presentation method, and make the delineated unit more suitable for the actual situation of the research area.

EUs are based on hydrological response units (HRUs); HRUs were originally defined as the basic units of simulation in the SWAT model [39]. HRUs conform to the fitness index method and are used to construct an adaptive index threshold system which can better delineate the ecosystem service zoning through hydrological analysis on the basis of DEM. The construction of HRUs is based on the hydrological analysis function in the ARCGIS software, and its theoretical basis is the D8 single flow algorithm [40]. However, the HRU divided by DEM can only represent one characteristic of terrain in space. There are also multiple types of land in the same unit, and the ecosystem services provided by them are also different. Therefore, the data on the soil type and vegetation type are used to divide the HRUs again, and the small patches of the boundary are merged to obtain the final EUs. The soil, vegetation, and watershed divisions of each ecological unit are different and have a certain uniqueness. The EUs constructed from this can better distinguish the boundaries of geographic units and reflect the mutual relationship between the ecosystem services.

3.2.3. Ecosystem Service Synergy/Trade-Off Time and Space Analysis Method

Synergy/Trade-Off Analysis Method of Ecosystem Services in Space

After determining that EUs will be used as carriers for defining radiation, we need to find a way to quantitatively and qualitatively study the synergy/trade-off of the ecosystem services in space. Spatial autocorrelation can be applied to express the aggregation characteristics of variables in space in this study [41,42]. Spatial autocorrelation refers to the potential interdependence between observation data of some variables in the same distribution area [43]. Statistically, through correlation analysis, whether there is a correlation between the changes of two phenomena can be detected. This is the same attribute variable of different observation objects, which is called autocorrelation. The parameters describing spatial autocorrelation include global autocorrelation and local autocorrelation. This study mainly uses bivariate local autocorrelation to express the synergy/trade-off of the ecosystem services in space.

Local spatial autocorrelation is a correlation index that uses the average value of adjacent positions to measure the degree of correlation between a position variable and other variables [44]. The calculation formula is as follows:

$$\mathrm{I} = \frac{n(x_i - \overline{x}) \sum_{j=1}^{m} W_{ij}(x_j - \overline{x})}{\sum_{i=1}^{n} (x_i - \overline{x})^2} \tag{8}$$

In the formula, X_i is the attribute value in the i cell, X_j is the attribute value in the j cell, and W_{ij} is the weight matrix representing the topological relationship of the space elements; m is the number of EUs adjacent to the EU i.

The bivariate local autocorrelation is defined as:

$$I_{lm}^{p} = z_l^p \sum_{q=1}^{n} W_{pq} \times z_m^q \tag{9}$$

In the formula, $z_l^p = \frac{X_l^p - \overline{X_l}}{\sigma_l}$; $z_m^q = \frac{X_m^q - \overline{X_m}}{\sigma_m}$; X_l^p is the value of the ecosystem service l of the space unit p; X_m^q is the value of the ecosystem service m of the space unit q; $\overline{X_l}$ and $\overline{X_m}$ are the values of ecosystem services l and m, respectively; and σ_l and σ_m are the variances of ecosystem services l and m, respectively.

Time-Scale Ecosystem Service Synergy/Trade-Off Analysis Method

The synergy/trade-off of the ecosystem services at the time scale is based on the change of ecological service functions every five years. Through correlation analysis, the synergy/trade-off relationship between the ecosystem services in different time series is calculated.

Using the unit-by-unit analysis method, the correlation coefficient between the two sets of long-term series of ecosystem services is calculated, and the positive and negative correlations are used to judge the synergy/trade-off between the two ecosystem services [38]. The correlation coefficient calculation formula is as follows:

$$R = \frac{\sum(x-\overline{x})(y-\overline{y})}{\sqrt{\sum(x-\overline{x})^2\sum(y-\overline{y})^2}} \tag{10}$$

According to the null hypothesis test of the correlation coefficient, the T test is used to evaluate the significance of the relationship between the ecosystem services at a time scale. The T test formula is as follows:

$$T = \frac{R}{\sqrt{\frac{1-R^2}{n-2}}} \tag{11}$$

In the formula, R is the corresponding partial correlation coefficient, and n is the number of observation samples. When $|T| > T_{0.05}$, i.e., $P < 0.05$, the null hypothesis is rejected, and the correlation result is significant. When $|T| > T_{0.01}$, i.e., $P < 0.01$, the null hypothesis is rejected, the correlation result is extremely significant.

The synergy/trade-off relationship of the ecosystem services iat the time scale is affected by many factors in the natural environment. The natural attribute characteristics of EUs, such as vegetation type and soil type, are correlated with the ecosystem service synergy/trade-off relationship, and the influence of natural attribute characteristics on the evolution of the ecosystem service synergy/trade-off is analyzed. Meanwhile, by the constructing pattern index, the relationship between the spatial characteristics of EUs and the synergy/trade-off of the ecosystem services at time scales is clarified. Based on the theory of landscape pattern and the use of EUs as carriers, the pattern index is applied to the study of the correlation between the spatial characteristics of the EU and the synergy/trade-off of the ecosystem services.

The Fragstats software is used to calculate the pattern index [45–48]. The software has three different research scales: the landscape scale, type scale, and patch scale. It is more appropriate to use the patch scale to study the pattern characteristics of geographic units. Therefore, this study selects the AREA, PERIM, GYRATE, PARA, SHAPE, and other indices (Table 2) to evaluate the pattern characteristics at the patch scale in order to explore the impact of the EU pattern characteristics on the trade-off/synergy of the ecosystem services.

Table 2. Pattern index and its description.

Pattern Index	Full Name of the Index
AREA	Patch Area
PERIM	Patch Perimeter
GYRATE	Radius of Gyration
PARA	Perimeter-Area Ratio
SHAPE	Shape Index
FRAC	Fractal Dimension Index
CIRCLE	Related Circumscribing Circle
CONTIG	Contiguity Index

4. Results

4.1. The Spatial Distribution and Change Characteristics of Ecosystem Services

The calculation results of water retention in 2000, 2005, 2010, and 2015 are compared (Figure 3). In the entire study area, the spatial differentiation of the water retention is more obvious, and it can be found that the water retention is greatly affected by land types in all years. In the past two decades, the water retention in the three regions of the Great Khingan Mountains, Lesser Khingan Mountains, and Changbai Mountains was relatively high. Among them, the water retention capacity at the southern foot of Changbai Mountain is the strongest, and the water retention capacity can reach 400 mm. The water retention in the east of the Lesser Khingan Mountains and other areas surrounding Changbai Mountain is about 150 mm. The water retention of the Great Khingan Mountains is generally about 100 mm. These three areas are covered by large areas of vegetation, which are dominated by larch, but they also contain a variety of trees and shrubs. The diversification of plants is conducive to the maintenance and improvement of the soil environment, and the developed root system and a good soil environment are conducive to the maintenance of soil moisture. During each rainfall, a large amount of water is fixed in the soil by plants, and the canopy interception of dense trees and shrubs can effectively absorb part of the rainfall. Therefore, forest land can conserve a huge amount of water and play an important role in water retention. The water retention of the Songnen Plain in the middle of the study area is lower than 50 mm. The Songnen Plain is mainly a farming area. As a large area of farmland is adjacent to Songhua River and Nen River, this area requires a huge amount of water due to the intensive planting of crops. The local residents generally drill wells around the farmland for irrigation, and the huge annual water demand results in the lowest water retention in the study area in the large-scale farmland distribution area. The amount of water retention in 2000 was between 0–678.73 mm; the amount of water retention in 2005 was between 0–622.42 mm; the amount of water retention in 2010 was between 0–651.46 mm; and the amount of water retention in 2015 was between 0–685.02 mm. Over time, the overall change is small. However, from 2005 to 2015, the water retention in the Great Khingan Mountains increased slightly. Based on the above results, it can be concluded that in a short period of time, the amount of water retention will not fluctuate too much, but some areas, such as the Great Khingan Mountains, will show a certain growth trend.

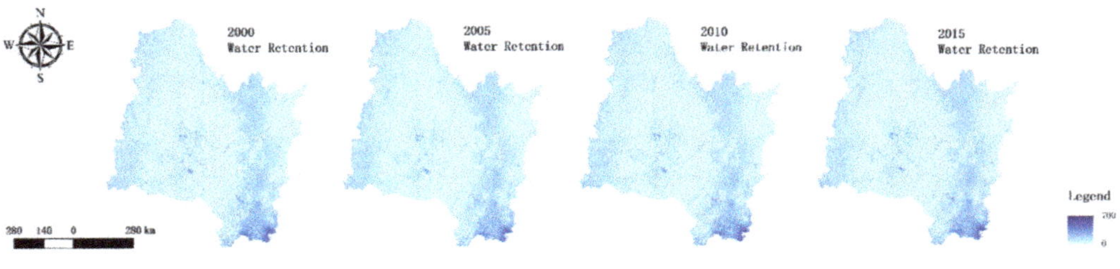

Figure 3. Distribution map of water retention.

The calculation results of soil conservation in 2000, 2005, 2010, and 2015 are compared (Figure 4). The spatial distribution characteristics of the soil conservation ecosystem services are similar to those of the water conservation ecosystem services. The three regions of the Great Khingan Mountains, Lesser Khingan Mountains, and Changbai Mountain in the northeast forest belt have higher values for soil conservation. The western side of the Lesser Khingan Mountains and Changbai Mountain shows a strong soil conservation ability. The amount of soil conservation in the Great Khingan Mountains is slightly lower than that of the other NFZ regions. Both the water retention and soil conservation in the study area are affected by the water and soil conservation function of the NFZ. A large area of arbor and shrub forest has the function of conserving water and soil at the same time. The Songnen Plain, like the middle and lower reaches of the Nenjiang River Basin and the Songhua River Secondary Basin, is a typical farming area, and the amount of soil conservation approaches zero. In 2000, the amount of soil conservation was between 0–2231.22 t/ha; in 2005, the amount of soil conservation was between 0–2335.92 t/ha; in 2010, the amount of soil conservation was between 0–2393.42 t/ha; and in 2015, the amount of soil conservation was between 0–2796.35 t/ha. Comparing the soil conservation capacity of the four years, the soil conservation capacity continued to increase from 2000 to 2015. There was a sudden increase in the soil conservation capacity in 2015, but the soil conservation capacity in the Greater Khingan Mountains area decreased. Based on the above results, it can be concluded that the soil retention capacity will show a slow growth trend. However, in certain areas, such as the Great Khingan Mountains, affected by factors such as vegetation, topography, and climate, the amount of soil conservation will fluctuate to a certain extent.

Figure 4. Distribution map of soil conservation.

The calculation results of carbon sequestration in 2000, 2005, 2010, and 2015 are compared (Figure 5). In the entire study area, the amount of carbon sequestration varies relatively smoothly in space, and the amount of carbon sequestration provided by the forest land is greater than the amount of carbon sequestration provided by the cultivated land. The amount of carbon sequestration in the NFZ is greater than that in other regions. From the analysis of the change characteristics of carbon sequestration in the past two decades, the amount of carbon sequestration in 2000 was between 117.80 and 1570.02 g/m^2; the amount of carbon sequestration in 2005 was between 119.39 and 1711.16 g/m^2; the amount of carbon sequestration in 2010 was between 129.60 and 1806.73 g/m^2; and the amount of carbon sequestration in 2015 was between 142.70 and 1650.73 g/m^2. From 2000 to 2010, the maximum amount of carbon sequestration continued to increase, reaching 1806.73 g/m^2 in 2010. Carbon sequestration services gather at the southern foot of Changbai Mountain, where the amount of carbon sequestration in this area increases at a rate of more than 100 g/m^2 every five years. The amount of carbon sequestration in the other regions showed a fluctuating trend. Among them, the amount of carbon sequestration in the Great Khingan Mountains and the Songnen Plain firstly increased and then decreased during the ten years from 2000 to 2010, while the amount of carbon sequestration in the Lesser Khingan Mountains increased slowly. In 2015, the maximum amount of carbon sequestration dropped to 1650.73 g/m^2. Except for the southern foot of Changbai Mountain, the amount of carbon sequestration in the other regions has shown an upward trend. Based on the above results, it can be concluded that the overall ecosystem services of carbon sequestration show an upward trend, but in individual years, the amount of carbon sequestration will fluctuate within a small range. Meanwhile, the amount of carbon sequestration at the southern foot of Changbai Mountain increased rapidly in the previous ten years and stabilized in a relatively high range in 2015.

The calculation results of biodiversity in 2000, 2005, 2010, and 2015 are compared (Figure 6). In the entire study area, the spatial differences in biodiversity are obvious, and biodiversity is greatly affected by land types in all years. In the past two decades, the biodiversity of the forest land in the Great Khingan Mountains, Lesser Khingan Mountains, and Changbai Mountain is far greater than that of the other land types. In 2000, 2005, 2010, and 2015, the value range of biodiversity is between 0–100. The biodiversity of the north-east forest belt has basically remained unchanged, and only some areas of the Songnen Plain show slight fluctuations. These areas are all areas with a small biodiversity base, and they cannot have a significant effect on the overall biodiversity. Based on the above results, it can be concluded that the biodiversity is relatively stable and does not frequently show huge changes. The forest land is more stable than the other land types, and the biodiversity of the forest land is much greater than that of the other land types.

Figure 5. Distribution map of carbon sequestration.

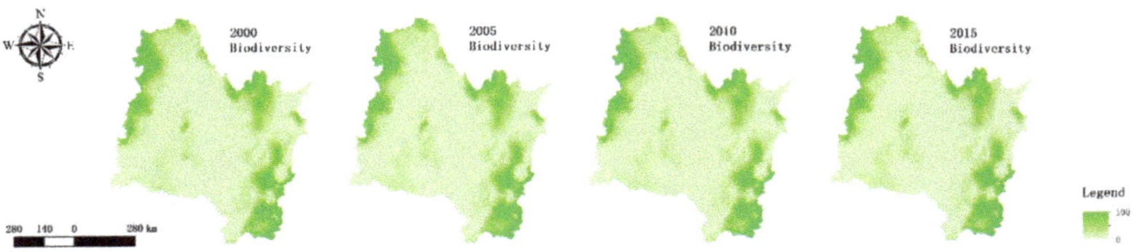

Figure 6. Distribution map of biodiversity.

The calculation results of crop production in 2000, 2005, 2010, and 2015 are compared (Figure 7). The spatial difference of regionalization in the study area is obvious. The areas with higher grain yields are mainly distributed in plain areas, such as the Songnen Plain and the lower Songhua River, among which the eastern side of the Songnen Plain is the highest. In the forest areas of the Great Khingan Mountains, Lesser Khingan Mountains, and Changbai Mountains, the crop production is almost 0. The crop production in 2000 was between 0 and 608.78 t/km^2; the crop production in 2005 was between 0 and 634.62 t/km^2; the crop production in 2010 was between 0 and 723.09 t/km^2; and the crop production in 2015 was between 0 and 783.63 t/km^2. A comparison of the crop production in the four years shows an upward trend year by year, and the crop production has continued to increase over the past two decades. The growth of crop production from 2000 to 2005 was relatively slow, and the g crop production from 2005 to 2015 showed rapid growth. Based on the above results, it can be concluded that the ecosystem service of crop production is relatively stable, and the increase of arable land and the update of farming technology have effectively increased the crop production, which is greatly affected by human factors.

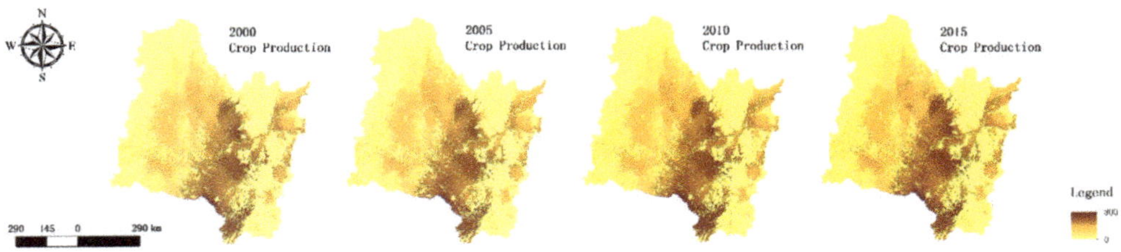

Figure 7. Distribution map of crop production.

4.2. Ecosystem Service Spatial Evaluation Carrier

According to the hydrological analysis using the ARCGIS software, based on DEM, 362 EUs are divided. However, the EUs are only divided according to the topography and rivers in the study area, and the spatial division of ecosystem services is also affected by vegetation and soil types. Therefore, the EUs based on the topography and rivers are combined with the soil vegetation types and soil types for secondary division. Finally, 2208 EUs are obtained. EUs are evaluation units with unique spatial location information and soil vegetation types. They are of great significance as carriers for presenting the synergy/trade-off relationship of ecosystem services in space. This is because the calculation results of ecosystem services are raster data, and in reality, ecosystem services are not bounded by a fixed grid unit size. EUs are more realistic response units, which are constructed based on actual conditions. Each EU has unique soil types, vegetation coverage, topographic slope characteristics, boundary characteristics, and spatial topological relationships. Therefore, it is highly scientific and feasible to use ecological units as carriers for ecosystem service synergy/trade-off analysis.

The EUs of the study area are shown in Figures 8 and 9, presenting the spatial distribution characteristics of the EUs according to the vegetation and soil types. The distribution of forests in the study area is the most concentrated. Among them, there are 671 EUs representing broadleaf forests, 149 EUs representing coniferous forests, and 40 EUs representing coniferous and broadleaf mixed forests. The forests are mainly distributed in the Great Khingan Mountains, Lesser Khingan Mountains, and Changbai Mountains. This combined area accounts for about 40% of the entire study area. The farmland is mainly distributed in the Songnen Plain. There is also a small part of farmland in the lower reaches of Songhua River. There are 829 EUs, and the area is similar to that of forest land. There are 339 EUs of grassland and meadows, which are scattered and interspersed between woodland and farmland and have no clustering characteristics. There are 72 EUs of shrubs, which are mainly distributed in the marginal areas of the Great Khingan Mountains and Changbai Mountains. There are 88 swamp EUs, which are concentrated among the forests of the Great Khingan Mountains. There are eleven types of soil in the SRZ, including chernozems, phaeozems, and luvisols. Among them, there are six types of chernozems, luvisols, phaeozems, cambisol, gleysols, and arenosols. The area of chernozems and phaeozems constitutes close to half of the study area. There are 1004 EUs, which are mainly distributed in the Songnen Plain and are suitable for farmland cultivation. There are 829 luvisols EUs, which are mainly distributed in the forest areas of the Great Khingan Mountains and Changbai Mountains. Cambisol, gleysols, and arenosols are scattered, but the area of a single patch is larger. Cambisol and gleysols are mainly distributed in forest land, while arenosols are interspersed in black soil fields.

4.3. Ecosystem Service Synergy/Trade-Off Relationship within the Spatial Area

The synergy/trade-off relationship between water retention and soil conservation in spatial from 2000 to 2015 is shown in Figure 10 and Table S1. Water retention and soil conservation are in a synergistic relationship on the whole. The synergy is mainly concentrated in four areas: the Songnen Plain, Changbai Mountain, the western side of the Lesser Khingan Mountains, and near the outlet of Songhua River. The Songnen Plain is the largest synergy gathering area, and its core area embodies an extremely significant synergy and shows a very significant synergistic relationship. This area is mainly a farming area, and the soil types are chernozems and phaeozems. The fringe area of the Songnen Plain is dominated by grassland and shrubs, which are shown to be in a significant synergistic relationship. However, the Songnen Plain has poor water retention and soil conservation capabilities. While it reflects a large area of synergy, the value of the overall ecosystem services is not large. The synergistic situation in the lower part of Songhua River is similar to that of the Songnen Plain, and the area is small. Both the western side of the Lesser Khingan Mountains and Changbai Mountain show a significant synergistic relationship between water retention and soil conservation. The Changbai Mountain shows a particularly significant synergistic relationship. The area is dominated by broadleaf forests. The soil types are luvisols and cambisol. A large-scale forest coverage effectively ensures that the water retention and soil conservation are maintained at a high level, reflecting synergy with a high aggregation. There are few EUs that reflect the trade-off relationship between water retention and soil conservation, and they are mainly distributed in the Songnen Plain marshes and some areas of meadows. The synergy/trade-off relationship between water retention and soil conservation in 2000, 2005, 2010, and 2015 is compared. The results in 2000, 2005, and 2010 show relatively consistent characteristics. Only in 2005, the synergistic relationship in the marginal area of the Songnen Plain fluctuated slightly. The stability and aggregation characteristics of the synergistic relationship in the NFZ were higher than those in the Songnen Plain. In 2015, compared with other years, the EUs with trade-off characteristics increased on a large scale. The water retention and soil conservation of the Songnen Plain were lower. Compared with the forest area, the stability of the cooperative relationship in the farming area was poor.

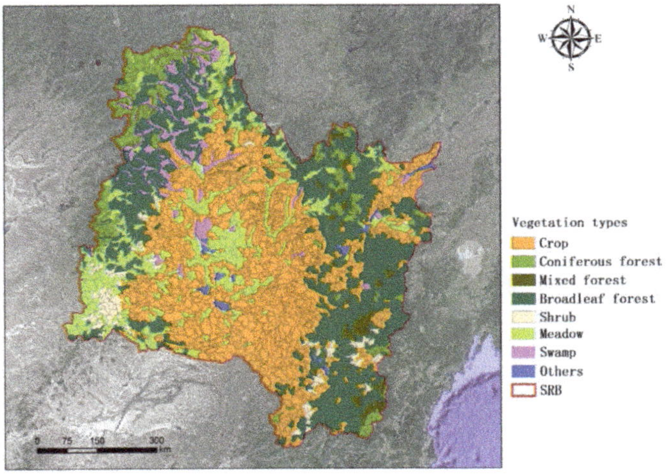

Figure 8. Distribution map of vegetation types.

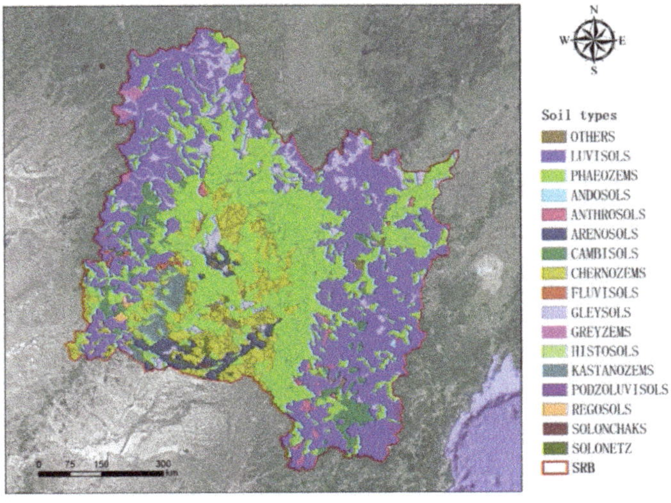

Figure 9. Distribution map of soil types.

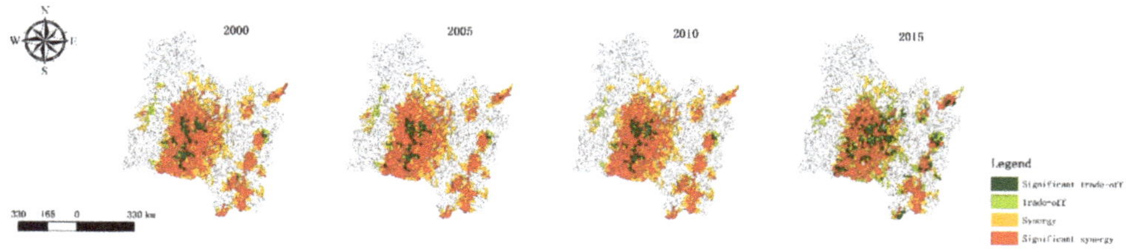

Figure 10. The synergy/trade-off relationship between water retention and soil conservation.

The synergy/trade-off relationship between water retention and biodiversity in spatial from 2000 to 2015 is shown in Figure 11 and Table S1. Water retention and biodiversity are in a synergistic relationship on the whole. The synergy is mainly concentrated in four areas: the northwestern part of the Great Khingan Mountains, the fringe area of the Songnen Plain, the eastern side of the Lesser Khingan Mountains, and the Changbai Mountains. The Great Khingan Mountains, Lesser Khingan Mountains, and Changbai Mountains are the components of the northeastern forest belt. The large-scale forest coverage has a high biodiversity level, and the water conservation of the northeast forest belt is maintained at a high level. Therefore, the water retention and biodiversity in the NFZ show a significant synergistic relationship. The Songnen Plain's farming areas have a poor biodiversity, and the water retention is much lower than that in the NFZ. Therefore, in the fringe area of the Songnen Plain, there are large areas of low-value gathering areas. In this area, the water retention and biodiversity present a significant synergistic relationship. The years 2000, 2005, and 2010 showed more consistent characteristics. In 2015, there was a very significant trade-off relationship in the northwestern part of the Greater Khingan Mountains. This indicates that there is a limit threshold for the synergy between water retention and biodiversity. When a certain ecosystem service exceeds the threshold, the synergy between the water conservation and biodiversity will immediately be reversed and become a trade-off relationship. Meanwhile, in 2015, a large number of EUs with trade-offs occurred at the edge of the Songnen Plain. In the cultivated land with a low water retention and biodiversity, the stability of the synergy and trade-off relationship between the ecosystem services is also poor.

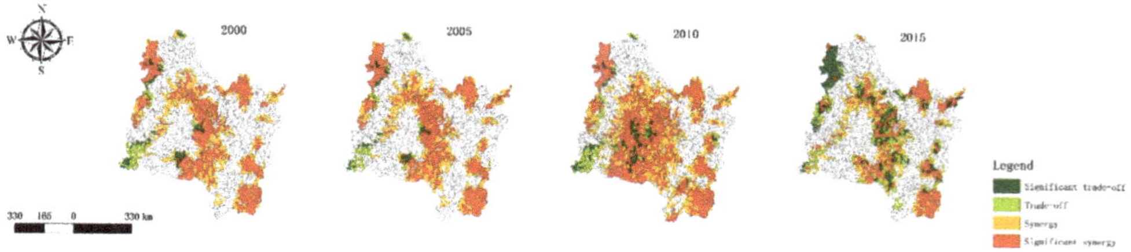

Figure 11. The synergy/trade-off relationship between water retention and biodiversity.

The synergy/trade-off relationship between the water retention and carbon sequestration at the spatial scale from 2000 to 2015 is shown in Figure 12 and Table S1. In 2000, the water retention and carbon sequestration showed a significant synergistic relationship in the core area of Changbai Mountain. The area was dominated by broadleaf forests and coniferous forests, and both the water retention and carbon sequestration reached high levels. While the meadows and arable land in the marginal area of Changbai Mountain increased, the water retention was greatly reduced, but the amount of carbon sequestration was maintained at a relatively high range in the arable land. Therefore, the marginal area of Changbai Mountain presents a trade-off relationship between water retention and carbon sequestration. There is a large area of significant synergy in the core area of the Songnen Plain, but there is a partial trade-off area on the western side. The grasslands and shrubs have a certain water retention capacity, but the amount of carbon sequestration is low. Therefore, it is embodied in the trade-off relationship between the water retention and carbon sequestration. In 2005, the significant synergy area of Changbai Mountain was reduced to the southern part of Changbai Mountain, the synergy area and trade-off area of the Songnen Plain were also reduced, and part of the synergy/trade-off area appeared in the northwest of the Great Khingan Mountains. In 2010, the two core areas of synergy/trade-off between the Songnen Plain and the southern foot of Changbai Mountain were still retained, and the synergy/trade-off area in the northwest of the Greater Khingan Mountains disappeared. Between 2000 and 2010, the core area of the Songnen Plain and the southern foot of Changbai Mountain were stable and significant synergistic areas. Meanwhile, some EUs at the edges of the two areas exhibit a trade-off relationship between water retention and carbon sequestration. In 2015, the Songnen Plain's synergy/trade-off area was reduced, and only the southern foot of Changbai Mountain remained a large-scale synergy/trade-off area. Comparing the changes over the past two decades, the synergy/trade-off area of water retention and carbon sequestration has gradually shrunk and finally converged on the southern side of Changbai Mountain, showing an extremely significant synergistic relationship between water retention and carbon sequestration.

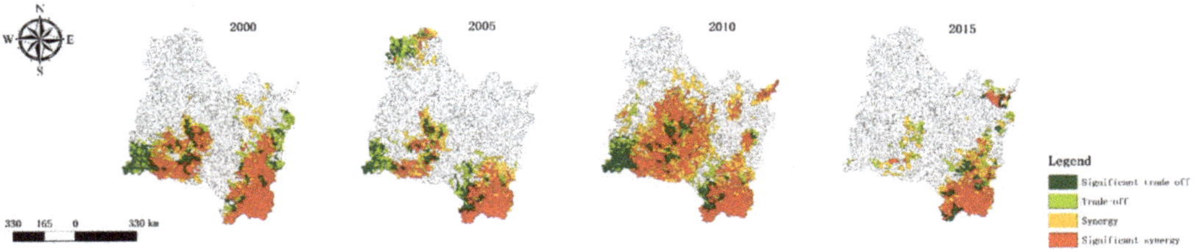

Figure 12. The synergy/trade-off relationship between water retention and carbon sequestration.

The synergy/trade-off relationship between soil conservation and biodiversity in the spatial characteristic from 2000 to 2015 is shown in Figure 13 and Table S1. The overall relationship between soil conservation and biodiversity is synergistic, and there has been little change over the past two decades. The northwestern part of the Great Khingan Mountains, parts of the Lesser Khingan Mountains, and Changbai Mountain show significant synergistic relationships. The northern and eastern marginal areas of the Songnen Plain also show a synergistic relationship, but the ecosystem service value is low.

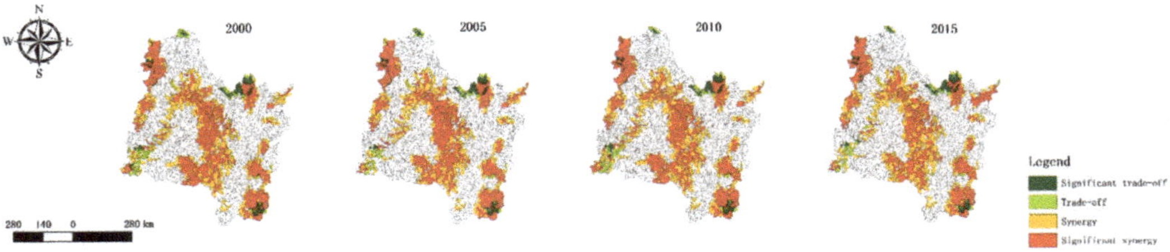

Figure 13. The synergy/trade-off relationship between soil conservation and biodiversity.

The synergy/trade-off relationship between soil conservation and carbon sequestration in the spatial characteristic from 2000 to 2015 is shown in Figure 14 and Table S1. The synergy/trade-off relationship between soil conservation and carbon sequestration is similar to the evolution of the distribution characteristics of the synergy/trade-off relationship between water retention and carbon sequestration. In the past two decades, the southern foot of Changbai Mountain has always been a very significant synergy zone. The synergy of the Songnen Plain, as a low-value synergy zone with a low soil conservation and carbon sequestration, has gradually weakened. The marginal area of the two synergistic core areas is partially expressed as an EU indicating a trade-off relationship.

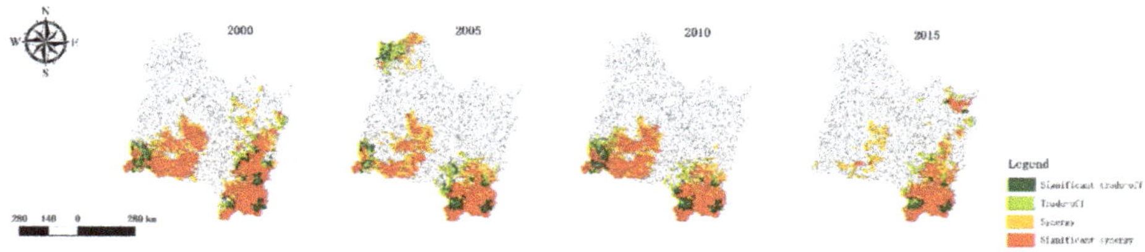

Figure 14. The synergy/trade-off relationship between soil conservation and carbon sequestration.

The synergy/trade-off relationship between biodiversity and carbon sequestration at the spatial scale from 2000 to 2015 is shown in Figure 15 and Table S1. In 2000, the biodiversity and carbon sequestration showed a significant synergistic relationship in the Songnen Plain. Meanwhile, there is a very significant synergy in the core area of Changbai Mountain. When the amount of carbon sequestration and biodiversity are maintained at a high/low level, the two show a significant synergy. At the western edge of Changbai Mountain, the biodiversity and carbon sequestration show a trade-off relationship. In 2005, the synergistic area of the Songnen Plain and Changbai Mountain was reduced. Part of the synergy/trade-off area appears in the northwest of the Great Khingan Mountains, and this area mainly shows a trade-off relationship. The synergy/trade-off relationship between biodiversity and carbon sequestration in 2010 is similar to that in 2000, but the synergistic area of Changbai Mountain has decreased. Compared with previous years, the synergy/trade-off relationship in 2015 had undergone major changes. The synergistic area of the Songnen Plain has been drastically reduced. Meanwhile, a large number of trade-off areas have been generated on the western side of Changbai Mountain, and the EUs showing a trade-off relationship have rapidly increased.

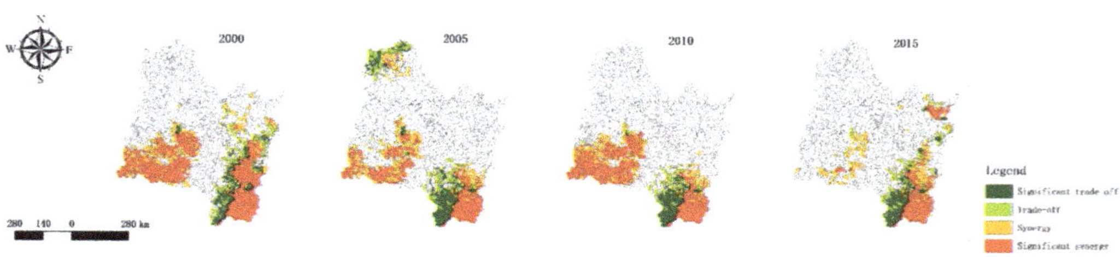

Figure 15. The synergy/trade-off relationship between carbon sequestration and biodiversity.

The synergy/trade-off relationship between crop production and water retention in the spatial characteristic from 2000 to 2015 is shown in Figure 16 and Table S1. Overall, there is a trade-off relationship between crop production and water retention between 2000 and 2010. The trade-off relationship is mainly distributed in four regions: the northern and eastern parts of the Songnen Plain, the lower Songhua River, the southern area of Changbai Mountain, and the eastern part of the Lesser Khingan Mountains. The Songnen Plain and the lower reaches of Songhua River are dominated by arable land, and the soil types are mainly phaeozems and chernozems. The characteristics of the trade-off between crop production and water retention in this area show that the crop production has a high-value accumulation and the water retention is low, with the two reflecting a trade-off relationship. The trade-off areas of Changbai Mountain and the Lesser Khingan Mountains are dominated by coniferous forests and broadleaf forests, and the soil types are luvisols, cambisols, and gleysols. The characteristics of the trade-off between crop production and water retention in this area are characterized by a low crop production and high water retention, and the two reflect a trade-off relationship. Changbai Mountain is dominated by broadleaf forests, showing a very significant trade-off relationship, while the coniferous forests and broadleaf forests in the eastern part of the Lesser Khingan Mountains are mixed, showing a relatively significant trade-off relationship. In 2015, the synergy/trade-off relationship between crop production and water retention had undergone major changes. The Songnen Plain area does not show any synergy/trade-off relationship. The northern part of the Great Khingan Mountains presents a significant synergistic relationship. The vegetation types are more complex, including coniferous forests, broadleaf forests, meadows, and swamps. The soil types are mainly luvisols. The synergistic feature is manifested in the low water retention and crop production. The southern part of Changbai Mountain presents a very significant synergistic relationship. The water retention in this area is relatively high, but the crop production is low. The western side of the area is a crop planting area, showing a partially synergistic relationship.

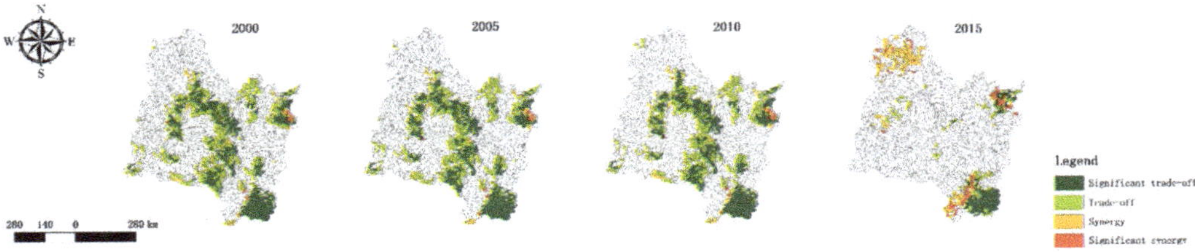

Figure 16. The synergy/trade-off relationship between water retention and crop production.

The synergy/trade-off relationship between crop production and soil conservation at the spatial scale from 2000 to 2015 is shown in Figure 17 and Table S1. The trade-off/synergy relationship in the past two decades is similar, and overall, there is a trade-off relationship between crop production and soil conservation. The trade-off relationship is mainly distributed in four regions: the Songnen Plain, the lower Songhua River, the eastern part of the Lesser Khingan Mountains, and Changbai Mountain. The Songnen Plain is the largest trade-off area. This area is dominated by crops, and the soil types are phaeozems and chernozems. The trade-off feature is embodied in the high crop production and low soil conservation. The trade-off characteristics of the lower Songhua River area are similar to those of the Songnen Plain. The trade-off area of the Lesser Khingan Mountains and Changbai Mountains is dominated by broadleaf forests and mixed forests. The soil types are luvisols and cambisols. The trade-off is characterized by a higher soil conservation and lower crop production. Some EUs in the core area of the Songnen Plain show a synergistic relationship. This area is mainly composed of meadows and swamps, and the crop production is low. Therefore, there is a synergistic relationship with soil conservation.

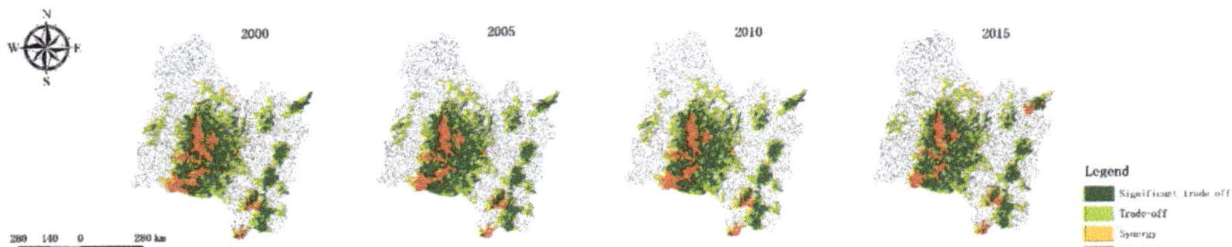

Figure 17. The synergy/trade-off relationship between crop production and soil conservation.

The synergy/trade-off relationship between crop production and biodiversity at the spatial scale from 2000 to 2015 is shown in Figure 18 and Table S1. The trade-off/synergy relationship in the past two decades is similar, and overall, there is a trade-off relationship between crop production and biodiversity. The trade-off relationship is mainly distributed in five regions: the Songnen Plain, the lower Songhua River, the western part of the Great Khingan Mountains, the eastern part of the Lesser Khingan Mountains, and Changbai Mountain. The eastern part of the Songnen Plain and the lower Songhua River have similar trade-off characteristics. The trade-off characteristics of the crop production and biodiversity in this area are characterized by a high value accumulation of crop production and low biodiversity, showing a trade-off relationship. The Great Khingan Mountains, Lesser Khingan Mountains, and Changbai Mountains also show a significant trade-off relationship. The vegetation types are mainly broadleaf forests, coniferous forests, and mixed forests, and the soil types are mainly luvisols. The trade-off is characterized by a high biodiversity and low crop production. There are still sporadic EUs in the Great Khingan Mountains and the Songnen Plain, showing a synergistic relationship. This area is mainly composed of meadows and shrubs, and the crop production is low. Therefore, there is a synergistic relationship with biodiversity.

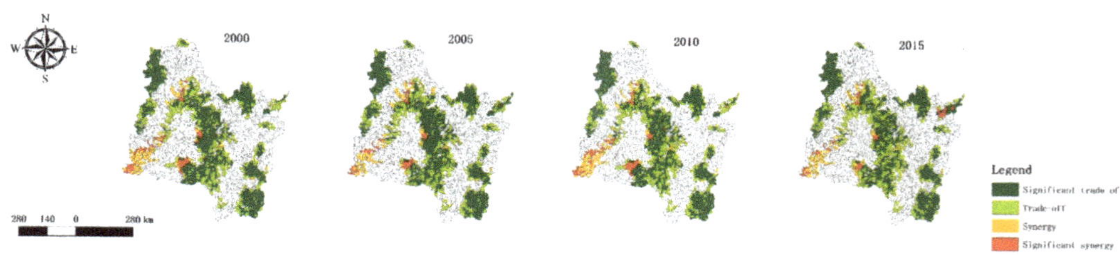

Figure 18. The synergy/trade-off relationship between crop production and biodiversity.

The synergy/trade-off relationship between crop production and carbon sequestration in the spatial charactaristic from 2000 to 2015 is shown in Figure 19 and Table S1. In 2000, crop production and carbon sequestration showed a very significant trade-off relationship in the core area of Changbai Mountain, which is dominated by broadleaf forests and coniferous forests. The characteristic of the trade-off is that the crop production is low, and the amount of carbon sequestration reaches a higher level. The meadows and arable land in the marginal area of Changbai Mountain have increased, and the crop production has increased significantly. The amount of carbon sequestration in the arable land has also remained within a relatively high range. Therefore, the marginal area of Changbai Mountain shows a synergistic relationship between crop production and carbon sequestration. There are some significant trade-off areas in the core area of the Songnen Plain, but there are some synergistic areas on the western side of the Songnen Plain, and the crop production and the amount of carbon sequestration both show low levels. In 2005, the significant trade-off area of Changbai Mountain was reduced to the southern part of Changbai Mountain, and the synergy/trade-off area of the Songnen Plain was also reduced. In the northwestern part of the Great Khingan Mountains, a synergistic area appears. This area is dominated by coniferous forests, broadleaf forests, and swamps, and the soil type is luvisols. The synergistic feature is reflected within the low range of crop production and carbon sequestration in this area. In 2010, the two core areas of the Songnen Plain and the southern foot of Changbai Mountain were still retained, and the synergistic area in the northwest of the Great Khingan Mountains disappeared. From 2000 to 2010, the southern foot of Changbai Mountain was very stable as a significant trade-off area, and some EUs at the edge showed a synergistic relationship between crop production and carbon sequestration. Meanwhile, the synergy/trade-off area between the western part of the Songnen Plain and the southern part of the Great Khingan Mountains is also relatively stable. In 2015, the Songnen Plain's synergy/trade-off area was greatly reduced, and only the southern foot of Changbai Mountain remained a large-scale synergy/trade-off area. Comparing the changes over the past two decades, the synergy/trade-off area between crop production and carbon sequestration has gradually shrunk and finally gathered on the southern side of Changbai Mountain, showing a significant trade-off relationship between crop production and carbon sequestration.

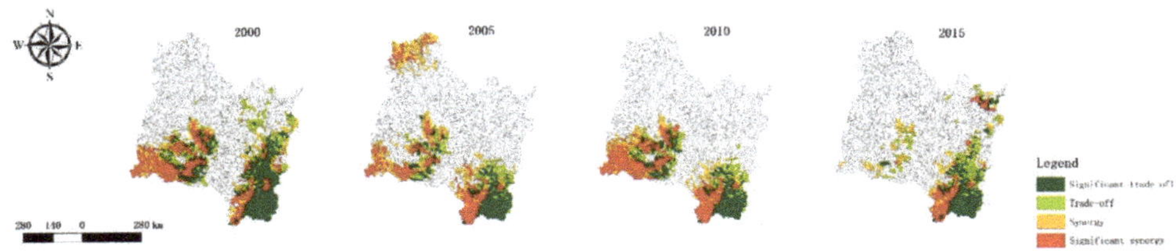

Figure 19. The synergy/trade-off relationship between crop production and carbon sequestration.

4.4. Ecological Service Function Time-Scale Synergy/Trade-Off Relationship

The synergy/trade-off relationship between water retention and soil conservation at the time scale from 2000 to 2015 is shown in Figure 20 and Table S2. The regions with a synergy/trade-off relationship are distributed in the Great Khingan Mountains, Lesser Khingan Mountains, and Changbai Mountain. The water retention and soil conservation are mainly synergistic. The 111 EUs present a significant synergistic relationship. The vegetation types are mainly shrubs and coniferous forests, as well as some meadows, shrubs, and planted crops. The soil types are mainly luvisols and phaeozems. The synergistic EUs are distributed more evenly in the NFZ. Relatively speaking, there are only 34 EUs showing a trade-off relationship, which are mainly distributed in the Lesser Khingan Mountains and Changbai Mountains. The soil types are mainly luvisols, and coniferous forests and broadleaf forests are the main vegetation types. The synergy/trade-off relationship between water retention and biodiversity at the time scale from 2000 to 2015 is shown in Figure 20 and Table S2. There are more EUs that are synergistic than trade-off EUs. Among them, there are some synergistic EUs in the Songnen Plain. In addition, the Northern Great Khingan Mountains, the eastern side of the Lesser Khingan Mountains, and the Changbai Mountains are intersected with synergy and trade-off regions. The synergistic EUs are mainly broadleaf forests, coniferous forests, and planted crops, and the soil types are mainly luvisols and phaeozems. The distribution of EUs showing a trade-off relationship is relatively scattered, but they are all contained in the NFZ. The vegetation types are mainly broadleaf forests, and the soil types include luvisols and phaeozems. The synergy/trade-off relationship between water retention and carbon sequestration at the time scale from 2000 to 2015 is shown in Figure 20 and Table S2. The areas in the study area showing the synergy/trade-off relationship are mainly distributed in the northern part of the Great Khingan Mountains, the western part of the Lesser Khingan Mountains, and the northern part of Changbai Mountain, with the most concentrated distribution in the north of Changbai Mountain. This area is a mixed area of planted crops and broadleaf forest. The altitude is not high, and the amount of water retention and carbon sequestration show a steady increase. On the whole, there are far more EUs showing a synergistic relationship than trade-off EUs, and 81 EUs showing a synergistic relationship are mainly distributed in the northern part of Changbai Mountain. This area is mainly constituted by broadleaf forests and planted crops, and the soil type is mainly luvisols. The synergy/trade-off relationship between soil conservation and biodiversity at the time scale from 2000 to 2015 is shown in Figure 20 and Table S2. EUs showing a synergy/trade-off relationship are mainly distributed in the NFZ. Among them, 84 EUs showed a significant synergistic relationship. Except for the NFZ, some farmland EUs in the Songnen Plain showed a synergistic relationship. The 45 EUs present a significant trade-off relationship, and the distribution is relatively scattered, with mainly broadleaf forests and coniferous forests, and the main soil type is luvisols. The synergy/trade-off relationship between soil conservation and carbon sequestration at the time scale from 2000 to 2015 is shown in Figure 20 and Table S2. EUs showing a synergy/trade-off relationship are gathered in the northern part of Changbai Mountain. There are more synergy EUs than trade-off EUs. The 69 synergy EUs are mainly broadleaf forests, and the soil types are mainly luvisols and phaeozems. There are 39 EUs showing a trade-off relationship, with mainly broadleaf forests and luvisols and phaeozems as the main soil types. The synergy/trade-off relationship between biodiversity and carbon sequestration at the time scale from 2000 to 2015 is shown in Figure 20 and Table S2. The biodiversity and carbon sequestration show a clear synergistic relationship at the time scale. Especially in the Songnen Plain, 474 EUs show a synergistic relationship, and planted crops and grassland meadows are the main vegetation types, indicating that biodiversity and carbon sequestration in the Songnen Plain both show a slow upward trend. The synergy/trade-off relationship between crop production and water retention at the time scale from 2000 to 2015 is shown in Figure 20 and Table S2. The crop production and water retention are embodied in a relatively scattered synergy/trade-off relationship on the time scale. Among them, the synergistic relationship is mainly distributed in the Great Khingan Mountains and Changbai Mountains. The vegetation types are mainly planted crops, broadleaf forests, and grasslands, and the soil types include chernozems and other types. The synergy feature is reflected in the increasing trend of the water retention and crop production. There are three trade-off ecological units in the Songnen Plain, and the trade-off relationship is reflected in the decrease in water retention and the increase in crop production. The synergy/trade-off relationship between the three types of ecosystem services of crop production and soil conservation, biodiversity, and carbon sequestration at the time scale from 2000 to 2015 is shown in the Figure 20 and Table S2. There is almost no synergy/trade-off relationship.

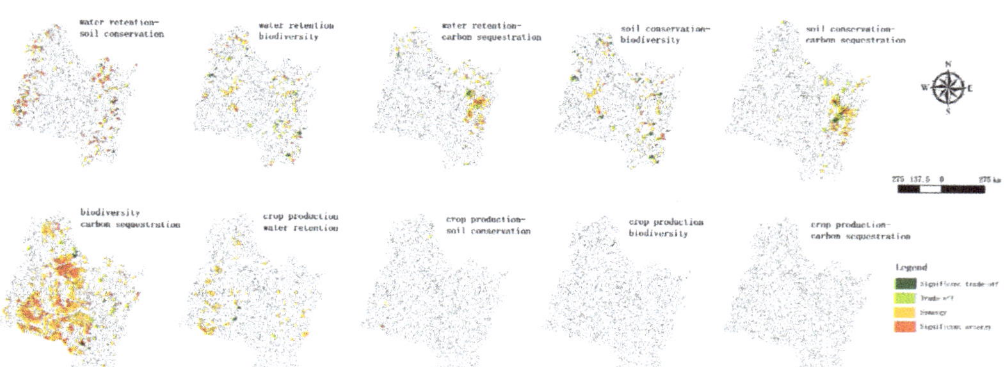

Figure 20. Time-scale ecosystem service synergy/trade-off.

4.5. Ecosystem Service Time-Scale Synergy/Trade-Off Spatial Topological Characteristics

Since EUs have unique spatial location characteristics, the pattern index is used to explore the relationship between ecosystem service synergy/trade-off and EU spatial location characteristics. The pattern index can describe the edge index characteristics and shape characteristics of the EUs. The AREA, PERIM, GYRATE, PARA, SHAPE, FRAC, CIRCLE, and CONTIG of all ecological units are calculated (Table 3). The values are compared with the corresponding parameters of the EUs that present a synergy/trade-off relationship. The comparison results are shown in the table. Compared with the average value, the AREA and PERIM of the EUs of the water retention and soil conservation synergy area show little difference, but the GYRATE is smaller than the overall average value. The AREA of the trade-off area and the GYRATE are both smaller than the overall average, and the PARA is slightly larger than the average. On the whole, the GYRATE of the EUs in the synergy/trade-off area of water retention and soil conservation is smaller than the average value, and the PARA is slightly larger than the average value. The AREA, PERIM, and GYRATE of the EUs in the synergistic area of water retention and biodiversity are smaller than the overall average, and the SHAPE and FRAC are also smaller than the average. This shows that the collaboration area is relatively regular, and the EUs are small. The GYRATE and the CIRCLE of the trade-off EUs are smaller than the average value, which indicates that the trade-off EUs are relatively round and smaller than the average. On the whole, the GYRATE of the EUs in the synergy/trade-off of water retention and biodiversity is smaller than the average value. The GYRATE, PERIM, SHAPE, and CIRCLE of the EUs in the synergistic area of water retention and carbon sequestration were less than the average value. The PERIM, GYRATE, SHAPE, and CIRCLE of the trade-off EUs are smaller than the average value, indicating that the trade-off area is more regular, and the EUs are smaller. On the whole, the EUs in the synergy/trade-off area of water retention and carbon sequestration are relatively regular, and the EUs are small. The GYRATE and PERIM of the EUs in the synergy/trade-off of soil conservation and biodiversity are smaller than the overall average, indicating that the EUs in the region are small. The AREA, PERIM, and GYRATE of the EUs in the synergistic zone of soil conservation and carbon sequestration are all smaller than the overall average, indicating that the synergy zone is small. On the whole, the synergistic relationship between soil conservation and carbon sequestration and the pattern index is more obvious. The EUs in the synergistic area of biodiversity and carbon sequestration are similar to the overall average. The trade-off is that the area of the regional ecological unit is larger than the overall average, and the PERIM, SHAPE, and CIRCLE are smaller than the average, indicating that the EUs are larger and more regular. The EUs in the synergistic area of crop production and water retention are similar to the overall average. There are too few synergy/trade-off areas between crop production and the other three types of ecosystem services to show a certain degree of regularity. In summary, the synergy/trade-off relationship of ecosystem services has a certain correlation with the spatial characteristics of the EUs. The AREA and PERIM of the EU showing the synergy/trade-off relationship are small, and the GYRATE is also significantly smaller than other EUs. Moreover, the shape of the EUs in these areas is more regular. The EU shape in the water retention and biodiversity synergy/trade-off area and soil conservation and carbon sequestration synergy/trade-off area is the most regular. The average EU area in the water retention and soil conservation synergy/trade-off area and the soil conservation and biodiversity synergy/trade-off area is the smallest.

Table 3. Ecosystem service synergy/trade-off pattern index comparison.

		COUNT	AREA	PERIM	GYRATE	PARA	SHAPE	FRAC	CIRCLE	CONTIG
Total		2208	255,370,298	130,851.9	7507.06	5.6874	2.0827	1.0745	0.6471	0.9596
water retention–soil conservation	synergy	111	227,606,158	124,243.2	7056.58	5.8289	2.0697	1.0736	0.6426	0.9585
	trade-off	34	212,045,203	125,661.8	6945.61	6.1833	2.1407	1.0769	0.6303	0.9561
water retention–biodiversity	synergy	89	204,312,638	108,196.6	6571.52	5.8172	1.9289	1.0683	0.6444	0.9586
	trade-off	42	245,070,447	123,869	6899.67	5.7379	2.0327	1.0727	0.624	0.9591
water retention–carbon sequestration	synergy	81	230,595,903	116,722.2	6775.22	5.5866	1.9632	1.0695	0.6179	0.9601
	trade-off	33	216,095,991	112,666.7	6559.17	5.6238	1.9488	1.0684	0.6028	0.9598
soil conservation–biodiversity	synergy	84	207,093,522	109,208.3	6568.3	5.8307	1.9354	1.0686	0.6381	0.9584
	trade-off	45	238,067,177	122,533.3	6909.94	5.8283	2.0357	1.0731	0.6322	0.9585
soil conservation–carbon sequestration	synergy	69	223,924,708	112,413	6682.68	5.5319	1.9147	1.0669	0.6076	0.9606
	trade-off	39	252,722,988	123,435.9	7109.12	5.262	1.9641	1.0694	0.6245	0.9624
biodiversity–carbon sequestration	synergy	474	251,662,399	130,127.6	7476.96	5.8463	2.099	1.075	0.6571	0.9584
	trade-off	31	276,115,852	126,177.4	7261.26	5.0286	1.9083	1.0659	0.6139	0.9641

Table 3. *Cont.*

		COUNT	AREA	PERIM	GYRATE	PARA	SHAPE	FRAC	CIRCLE	CONTIG
crop production–water retention	synergy	77	240,889,394	127,935.1	7353.74	5.7928	2.0886	1.0748	0.6607	0.9588
	trade-off	3	357,468,795	149,666.7	8482.68	4.2188	1.9696	1.0687	0.6133	0.9701
crop production–soil conservation	synergy	2	230,590,587	146,000	8757.33	6.266	2.3735	1.0892	0.7343	0.9552
	trade-off	1	276,198,068	153,500	8160.01	5.549	2.291	1.086	0.7508	0.9606
crop production–biodiversity	synergy	0	0	0	0	0	0	0	0	0
	trade-off	1	128,654,449	86,000	6267.28	6.6764	1.8901	1.0684	0.7508	0.9525
crop production–carbon sequestration	synergy	0	0	0	0	0	0	0	0	0
	trade-off	0	0	0	0	0	0	0	0	0

5. Discussion

The Songhua River basin in the study area mainly includes four large areas: the Great Khingan Mountains, Lesser Khingan Mountains, Songnen Plain, and Changbai Mountain, among which the Great Khingan Mountains, Lesser Khingan Mountains, and Changbai Mountain together constitute the NFZ in the National Ecological Barrier Zone. The distribution of the five types of ecosystem services, including water retention, soil conservation, biodiversity, carbon sequestration, and crop production, was significantly different between 2000 and 2015. Comparing the change characteristics of ecosystem services from 2000 to 2015, water conservation will not fluctuate much in the short term, but some areas, such as the Greater Xing'an Mountains, will show a certain growth trend: soil conservation shows a trend of slow growth, but in certain areas, such as the Greater Khingan Mountains, affected by factors such as vegetation, topography, and climate, the amount of soil conservation will fluctuate to a certain extent; carbon sequestration show an upward trend as a whole, but in individual years, the amount of carbon fixation and oxygen release will fluctuate in a small range; biodiversity is relatively stable; and crop production is relatively stable, and the increase in arable land and the upgrading of farming technology have effectively increased grain production, which is greatly affected by human factors. The NFZ is dominated by broadleaf forests, coniferous forests, and mixed forests. A variety of broadleaf, coniferous, and shrub vegetation is widely distributed in this area. The diversification of plants is conducive to the maintenance and improvement of a soil environment, and a developed root system and good soil environment are conducive to the maintenance of soil moisture. During each rainfall, a large amount of water is fixed in the soil by plants, and the canopy interception of dense trees and shrubs can effectively absorb part of the precipitation. Therefore, forest land can conserve a huge amount of water, and at the same time has a high amount of soil conservation [49,50]. The amount of carbon sequestration is affected by NPP. Biodiversity is defined by the complexity of the species in the system. The NFZ is a primitive large-scale natural forest with high vegetation coverage and diverse types. Therefore, biodiversity and carbon sequestration are also maintained in a relatively high range [51–54]. The Songnen Plain, as the alluvial plain of Songhua River and Nenjiang River, is the main food production area in the northeast. There is a lot of cultivated land, mainly planted crops, and the soil type is chernozems, which is suitable for crop growth [55]. In this area, the crop production is relatively high, while the other four types of ecosystem services are relatively low.

The synergy/trade-off relationship of ecosystem services at the spatial scale is shown in Figure 21. Water retention has a strong correlation with the other four types of ecosystem services. Among them, water retention and crop production are in a trade-off relationship, and water retention and the other three types of ecosystem services are in a synergistic relationship. Yin believes that the synergy between water retention and soil conservation is an important synergy in the TEBTS of the National Barrier Zone in his research on the synergy/trade-off of ecosystem services in the National Barrier Zone [56]. Soil conservation is similar to water retention. It has a strong correlation with the other four types of ecosystem services. Among them, soil conservation and crop production are in a trade-off relationship, and the other three types of ecosystem services are in a synergistic relationship. Biodiversity is not highly related to carbon sequestration and is highly related to the other three types of ecosystem services. Biodiversity and crop production are in a trade-off relationship, and there is a synergistic relationship with the other three types of ecosystem services. Carbon sequestration is not highly correlated with the other ecosystem services. They are in a trade-off relationship with crop production, and they are in a synergistic relationship with the other three types of ecosystem services. Crop production and the other four types of ecosystem services all present a trade-off relationship. In his research on the synergy/trade-off of ecosystem services in Northeast China, Qi also found that the three types of ecosystem services, including crop production and water retention, are in a trade-off relationship, and water retention and soil conservation are in a significant synergistic relationship [57]. At the time scale, the synergy/trade-off relationship among the four types of ecosystem services, including water retention, soil conservation, biodiversity, and carbon sequestration, is mainly synergistic. The synergy/trade-off areas are mainly distributed in the NFZ, but the EUs showing a significant synergy/trade-off relationship are relatively scattered [58]. The EUs showing a synergistic relationship are mainly broadleaf forests, and the soil type is luvisols. Meanwhile, the ecosystem service synergy/trade-off relationship has a certain correlation with the spatial characteristics. The EUs with a synergy/trade-off relationship are smaller in the AREA and PERIM, and the GYRATE is also significantly smaller than that of the other EUs. Moreover, the shape of the EUs in this area is more regular. Compared with the other EUs, the boundary between this type of EU and the surrounding EUs is smaller and stable. Over a certain period of time, the probability of ecological energy interaction is reduced, so that the ecosystem services maintain a stable change trend and finally show a significant synergy/trade-off relationship. This shows that the four types of ecosystem services—water retention, soil conservation, biodiversity, and carbon sequestration—have a greater probability of showing a synergy/trade-off relationship under the conditions of these attributes and spatial characteristics [59,60]. Crop production and water retention are mainly synergistic at the time scale [56], and the synergy feature is reflected in the increase in water retention and crop production year by year. The synergy/trade-off relationship between crop production and water retention does not show any correlation with the spatial characteristics of EUs.

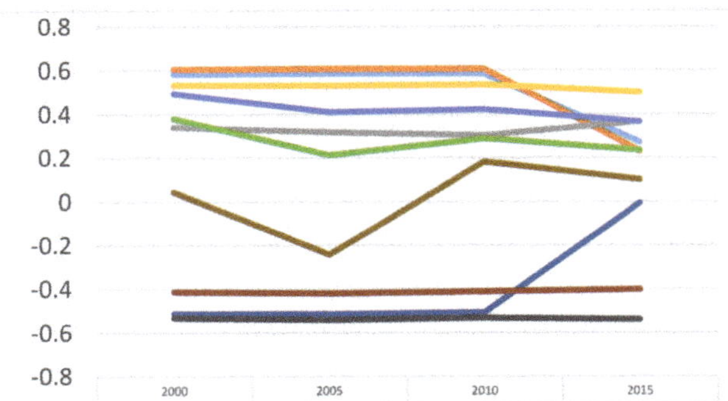

	2000	2005	2010	2015
water retention to soil conservation	0.581	0.584	0.585	0.272
water retention to biodiversity	0.603	0.606	0.606	0.23
water retention to carbon sequestration	0.341	0.317	0.303	0.367
soil conservation to biodiversity	0.532	0.529	0.533	0.499
soil conservation to carbon sequestration	0.494	0.41	0.422	0.365
biodiversity to carbon sequestration	0.38	0.212	0.288	0.233
crop production to water retention	-0.513	-0.515	-0.508	-0.007
crop production to soil conservation	-0.414	-0.421	-0.412	-0.403
crop production to biodiversity	-0.532	-0.541	-0.532	-0.54
crop production to carbon sequestration	0.043	-0.243	0.182	0.099

Figure 21. Ecosystem service synergy/trade-off change trend.

However, this study still has some limits, which are as follows. When analyzing the synergy/trade-off relationship between EU attributes and spatial characteristics for ecosystem services, only natural attributes, such as vegetation types and soil types, are considered, and other factors affecting ecosystem services are not considered. In recent years, human activities have had an increasing impact on ecosystem service synergy/trade-off relationships. In particular, countries in key ecological regions have formulated relevant ecological restoration or ecological optimization policies, which will seriously affect ecosystem service synergy/trade-off relationships. Therefore, the next step is to consider more natural and social attributes and incorporate them into the spatial database of synergy/trade-off analysis. EUs are carriers in the study of the time-scale ecosystem service synergy/trade-off relationship. Within a certain period of time, EUs may have small-scale changes. The use of unified EUs is beneficial to the research itself, but they may have slight differences from the actual situation. In addition, the results show that the impact of ecosystem services is not limited to EUs, as the change of a certain ecosystem service may cause the response of other ecosystem services within a certain range. Ecosystem services not only have internal effects, but also external benefits. When the "source" and "sink" of ecosystem services are not uniform in space, the phenomenon of ecosystem service flow occurs. Therefore, we will consider this phenomenon in further research exploring the relationship between different ecosystem services inside and outside EUs by simulating the ecosystem service flow.

This research explores the synergy/trade-off relationship of ecosystem services at the spatial and time scales by constructing EUs. The essence of ecosystem service synergy/trade-offs in the spatial characteristic is to express the spatial aggregation characteristics of ecosystem services, and the essence of time-scale ecosystem service synergy/trade-offs is to express the relationship between the current and future use of ecosystem services. Focusing on the relationship between ecosystem service synergy/trade-offs and attribute and spatial characteristics, the impact of attribute characteristics and spatial characteristics on ecosystem service synergy/trade-offs can be qualitatively analyzed. The influencing factors of ecosystem service synergy/trade-offs can be clarified, incorporating key areas with different ecosystem services into the ecosystem, integrating ecosystem services, and using this as a basis to optimize the ecosystem in order to maximize the value of regional ecosystem services [11,54]. The NFZ in this study has a higher water retention, soil conservation, and other regulating and supporting functions. Each barrier zone in the TEBTS has a primary ecosystem service, and the synergy/trade-off between the primary ecosystem service and the secondary ecosystem service is coordinated. This is of great significance for improving the overall benefits of ecosystem services, ensuring the sustainable supply of ecosystem services, and realizing the "win–win" of human society and ecosystems [11]. Meanwhile, exploring the synergy/trade-off relationship of ecosystem services at spatial and time scales will provide a certain data basis and theoretical support for major projects, such as ecological restoration and ecological environment governance, which is of great significance for the improvement of the ecological security pattern [56].

6. Conclusions

Comparing the ecosystem services from 2000 to 2015, the water retention is between 0 and 700 mm, showing the characteristics of fluctuations. The water retention of the Lesser Khingan Mountains and Changbai Mountains is relatively large. The amount of soil conservation is between 0 and 2800 t/ha. The soil conservation reached its maximum in 2015, and the ecosystem services of soil conservation are higher in the northeastern forest zone. The amount of carbon sequestration is between 100–1900 g/m^2. The amount of carbon sequestration in the northeastern forest zone is greater than that in other regions, but the spatial changes are relatively gentle, and the spatial difference is not obvious. From 2010 to 2015, the amount of carbon sequestration increased rapidly. The biodiversity is between 0–100, with obvious spatial differences. The forest areas, such as the Great Khingan Mountains, have much more biodiversity than other areas. The biodiversity is relatively stable, with almost no change in the past two decades. The crop production is between

0–800 t/km^2, showing an increasing trend year by year. As the main crop production area, the Songnen Plain provides the main crop production function, which is greatly affected by humans. The ecological units are used as the research carriers, and 2208 ecological units are delineated based on the topography, vegetation type, and soil type. The ecological units include eight types of vegetation, such as broadleaf forests, and eleven types of soil, such as luvisols. Not only can ecological units better define the spatial scope of ecosystem services, but they also have spatial topological characteristics and attribute characteristics. Based on this, the influence of spatial characteristics and attribute characteristics on the change trend of ecosystem service synergy/trade-off relationships is analyzed.

At the spatial scale, the four types of ecosystem services—water retention, soil conservation, biodiversity, and carbon sequestration—present a synergistic relationship. Among them, the synergistic relationships between water retention and soil conservation, between water retention and biodiversity, and between soil conservation and biodiversity are the most significant. Crop production and the other four types of ecosystem services all present a trade-off relationship in space. Among them, crop production and water retention and crop production and biodiversity present a very significant trade-off relationship, crop production and soil conservation present a trade-off relationship, and there is no significant relationship between crop production and carbon sequestration. At the time scale, there is a synergistic relationship among the four types of ecosystem services: water retention, soil conservation, biodiversity, and carbon sequestration. Crop production and water retention present a synergistic relationship, but there is no significant synergy/trade-off relationship with the other three types of ecosystem services. The vegetation types showing a synergy/trade-off relationship are mainly broadleaf forests, with some coniferous forests and crops. The soil types are dominated by luvisols and phaeozems. The ecological units showing a synergy/trade-off relationship are mainly distributed in the northeastern forest zone and have certain spatial topological characteristics: the EU showing a synergy/trade-off relationship is smaller in the AREA and PERIM, the GYRATE is also significantly smaller than that of the other EUs, and the shape of the EUs are more regular.

In this study, we focused on the spatial aggregation characteristics and change trends of ecosystem service synergy/trade-offs and clarified the influencing factors of ecosystem service synergy/trade-offs by analyzing the influence of various characteristics on the change trends. Based on this, we integrated ecosystem services, optimized the ecosystem, maximized the value of regional ecosystem services, and provided a data foundation and theoretical support for improving the ecological security pattern.

The application prospects of this research are prominently reflected: based on the ecological unit, the spatial boundary of ecosystem services is clarified, and the spatial characteristics and attribute characteristics of ecosystem services are analyzed; and simulating the dynamic characteristics of natural factors affecting ecosystem services and analyzing the characteristics of different spatial attributes leads to the synergy/trade-off of ecosystem services. In summary, this study makes the research on ecosystem service synergy/trade-off more diversified. The research carrier is upgraded from a regular grid to an ecological unit that expresses spatial and attribute characteristics. At the same time, the relationship between spatial characteristics and attribute characteristics and the synergy/trade-off of ecosystem services has been established, which lays the foundation for studying the mechanism of ecosystem services' synergy/trade-off response.

Supplementary Materials: The following are available online at https://www.mdpi.com/article/10.3390/f12080992/s1, Table S1. Space-scale ecosystem service synergy/trade-off. Table S2. Time-scale ecosystem service synergy/trade-off.

Author Contributions: Conceptualization, T.N. and J.Y.; methodology, T.N.; software, T.N.; validation, T.N. and D.Y.; formal analysis, L.Y.; investigation, Q.L.; resources, X.M.; data curation, Q.L.; writing—original draft preparation, Y.H.; writing—review and editing, T.N.; visualization, T.N.; supervision, D.Y.; project administration, D.Y.; funding acquisition, D.Y. All authors have read and agreed to the published version of the manuscript.

Funding: This research is supported by the National Key Research and Development Program of China (2016YFC0507303).

Acknowledgments: Thank you for the data support provided by the research group "National Scale Ecosystem Service Evaluation and Analysis Model System". The authors also thank the reviewers for their valuable suggestions that helped to improve the quality of the manuscript.

Conflicts of Interest: The authors declare no conflict of interest.

References

1. Qin, X.N.; Lu, X.L.; Wu, C.Y. The knowledge mapping of domestic ecological security research: Bibliometric analysis based on citespace. *Acta Ecol. Sin.* **2014**, *34*, 3693–3703.
2. Wang, Y.; Pan, J.H. Building ecological security patterns based on ecosystem services value reconstruction in an arid inland basin: A case study in Ganzhou District, NW China. *J. Clean. Prod.* **2019**, *241*, 118337. [CrossRef]
3. Wang, X.F.; Lesi, M.C.; Zhang, M.M. Ecosystem pattern change and its influencing factors of "two barriers and three belts". *Chin. J. Ecol.* **2019**, *38*, 2138–2148.
4. Wang, Y.; Li, X.M.; Zhang, F.; Wang, W.W.; Xiao, R.B. Effects of rapid urbanization on ecological functional vulnerability of the land system in Wuhan, China: A flow and stock perspective. *J. Clean. Prod.* **2019**, 119284. [CrossRef]
5. Agetsuma, N. Ecological function losses caused by monotonous land use induce crop raiding by wildlife on the island of Yakushima, southern Japan. *Ecol. Res.* **2007**, *22*, 390–402. [CrossRef]
6. Du, L.J.; Wang, X.R.; Liu, Y. Ecological compensate standard accounting of soil and water conservation. *J. Hydraul. Eng.* **2010**, *41*, 1346–1352.
7. Prober, S.M.; Thiele, K.R.; Lunt, I.D.; Koen, T.B. Restoring Ecological Function in Temperate Grassy Woodlands: Manipu-lating Soil Nutrients, Exotic Annuals and Native Perennial Grasses through Carbon Supplements and Spring Burns. *J. Appl. Ecol.* **2005**, *42*, 1073–1085. [CrossRef]
8. Yao, E.M.; Xiao, Y.; Ouyang, Z.Y.; Zheng, H. Spatial characteristics of soil conservation service and its impact factors in Hainan Island. *Acta Ecol. Sin.* **2013**, *33*, 83–92.
9. Bennett, E.M.; Peterson, G.D.; Gordon, L.J. Understanding relationships among multiple ecosystem services. *Ecol. Lett.* **2009**, *12*, 1394–1404. [CrossRef] [PubMed]
10. Li, S.; Wang, J.; Zhu, W.; Zhang, J.; Liu, Y.; Gao, Y.; Li, Y. Research framework of ecosystem services geography from spatial and regional perspectives. *Acta Geogr. Sin.* **2014**, *69*, 1628–1639.
11. Dai, E.; Wang, X.; Zhu, J.; Gao, J. Progress and Perspective on Ecosystem Services Trade-offs. *Prog. Geogr.* **2015**, *30*, 1250–1259.
12. Fu, B.; Yu, D. Trade-off analyses and synthetic integrated method of multiple ecosystem services. *Resour. Sci.* **2016**, *38*, 1–9.
13. Zhao, W.; Liu, Y.; Feng, Q.; Wang, Y.; Yang, S. Ecosystem services for coupled human and environment systems. *Prog. Geogr.* **2018**, *37*, 139–151.
14. Lautenbach, S.; Volk, M.; Whittaker, G.; Seppelt, R. Optimization-based trade-off analysis of biodiesel crop production for managing an agricultural catchment. *Environ. Model. Softw.* **2013**, *48*, 98–112. [CrossRef]
15. Oken, K.L.; Essington, T.E. Evaluating the effect of a selective piscivore fishery on rockfish recovery within marine pro-tected areas. *ICES J. Mar. Sci.* **2016**, *73*, 2267–2277. [CrossRef]
16. Vauhkonen, J.; Ruotsalainen, R. Assessing the provisioning of ecosystem services in a Scandinavian boreal forest: Suitability and tradeoff analyses on grid-based wall-to-wall forest inventory data. *For. Ecol. Manag.* **2017**, *389*, 272–284. [CrossRef]
17. White, C.; Halpern, B.C.; Kappel, C.V. Ecosystem service tradeoff analysis reveals the value of marine spatial planning for multiple ocean uses. *Proc. Natl. Acad. Sci. USA* **2012**, *109*, 4696–4701. [CrossRef] [PubMed]
18. Gissi, E.; Gaglio, M.; Reho, M. Sustainable energy potential from biomass through ecosystem services trade-off analysis: The case of the Province of Rovigo (Northern Italy). *Ecosyst. Serv.* **2016**, *18*, 1–19. [CrossRef]
19. Chan, K.M.A.; Shaw, M.R.; Cameron, D.R.; Underwood, E.C.; Daily, G.C. Conservation planning for ecosystem services. *PLoS Biol.* **2006**, *4*, 2138–2152. [CrossRef] [PubMed]
20. Egoh, B.; Reyers, B.; Rouget, M.; Bode, M.; Richardson, D.M. Spatial congruence between biodiversity and ecosystem services in South Africa. *Biol. Conserv.* **2009**, *142*, 553–562. [CrossRef]
21. Raudsepp-Hearne, C.; Peterson, G.D.; Bennett, E.M. Ecosystem service bundles for analyzing tradeoffs and services via functional traits. *Biodivers. Conserv.* **2010**, *19*, 2873–2893.
22. Chen, D.; Li, J.; Yang, X.; Liu, Y. Trade-offs and optimization among ecosystem services in the Weihe River basin. *Acta Ecol. Sin.* **2018**, *38*, 3260–3271.
23. Sun, Y.; Ren, Z.; Zhao, S.; Zhang, J. Spatial and temporal changing analysis of synergy and trade-off between ecosystem services in valley basins of Shaanxi Province. *Acta Geogr. Sin.* **2017**, *72*, 104–109.
24. Ning, J.; Liu, G.; Liu, Q.; Xie, C. Spatial discretization of hydrological response units and improved SWAT model. *Adv. Water Sci.* **2012**, *23*, 14–20.
25. Khan, U.; Tuteja, N.K.; Sharma, A. Delineating hydrologic response units in large upland catchments and its evaluation using soil moisture simulations. *Environ. Model. Softw.* **2013**, *46*, 142–154. [CrossRef]

26. Miao, C.; Yang, L.; Liu, B.; Gao, Y.; Li, S.L. Streamflow changes and its influencing factors in the mainstream of the Songhua River Basin, Northeast China over the past 50 years. *Environ. Earth Sci.* **2010**, *63*, 489–499. [CrossRef]
27. Yang, Y.H.; Yan, B.; Shen, W. Assessment of point and nonpoint sources pollution in Songhua River Basin, Northeast China by using revised water quality model. *Chin. Geogr. Sci.* **2010**, *20*, 30–36. [CrossRef]
28. Li, Y.; Xu, L.Y.; Li, S. Water Quality Analysis of the Songhua River Basin Using Multivariate Techniques. *J. Water Res. Prot.* **2009**, *1*, 110–121. [CrossRef]
29. Gong, S.; Xiao, Y.; Zheng, H.; Xiao, Y.; Ouyang, Y. Spatial patterns of ecosystem water conservation in China and its impact factors analysis. *Acta Ecol. Sin.* **2017**, *37*, 2455–2462.
30. Wang, Z.; Zhang, K.; Xue, X.; Hong, Z.; Hong, Y.; Gourley, J. Water balance-based actual evapotranspiration reconstruction from ground and satellite observations over the conterminous United States. *Water Resour. Res.* **2015**, *51*, 6485–6499.
31. Shima, N.; Hossein, A.; Ali, N. Simulation of water balance equation components using SWAT model in Samalqan Water-shed (Iran). *Arab. J. Geosci.* **2020**, *13*, 1–15.
32. Ranzi, R.; Le, T.H.; Rulli, M.C. A RUSLE approach to model suspended sediment load in the Lo river (Vietnam): Effects of reservoirs and land use changes. *J. Hydrol.* **2012**, *422*, 17–29. [CrossRef]
33. Wu, L.; Yan, L. Impact of road on soil erosion risk pattern based on RUSLE and GIS: A case study of Hangjinqu highway, Zhuji section. *Acta Ecol. Sin.* **2014**, *34*, 5659–5669.
34. Chen, S.; Yang, X.; Xiao, L.; Cai, H. Study of Soil Erosion in the Southern Hillside Area of China Based on RUSLE Model. *Resour. Sci.* **2014**, *36*, 1288–1297.
35. Hazem, A.; Juliet, S. Mapping the soil loss in Marqya basin: Syria using RUSLE model in GIS and RS techniques. *Environ. Earth Sci.* **2017**, *76*, 1–10.
36. Yang, X. Deriving RUSLE cover factor from time-series fractional vegetation cover for hillslope erosion modelling in New South Wales. *Soil Res.* **2014**, *52*, 253–261. [CrossRef]
37. Onori, F.; Bonis, P.; Grauso, S. Soil erosion prediction at the basin scale using the revised universal soil loss equation (RUSLE) in a catchment of Sicily (southern Italy). *Environ. Geol.* **2006**, *50*, 1129–1140. [CrossRef]
38. Li, L.; Wu, D.; Wang, F.; Liu, Y.; Liu, Y.; Qian, L. Prediction and tradeoff analysis of ecosystem service value in the rapidly urbanizing Foshan City of China: A case study. *Acta Ecol. Sin.* **2020**, *40*, 9023–9036.
39. Jayakrishnan, R.; Srinivasan, R.; Santhi, C.; Arnold, J.G. Advances in the Application of the SWAT Model for Water Resources Management. *Hydrol. Process.* **2005**, *19*, 749–762. [CrossRef]
40. Yan, Z.; Ming, Y.; Roeder, E.; Hicks, R.W.; Yang, J.Z. Estimating Ammonium and Nitrate Load from Septic Systems to Sur-face Water Bodies within ArcGIS Environments. *J. Hydrol.* **2015**, *532*, 177–192.
41. Koenig, W.D. Spatial autocorrelation of ecological phenomena. *Trends Ecol. Evol.* **1999**, *14*, 22–26. [CrossRef]
42. Zhang, L.J.; Ma, Z.H.; Luo, G. An Evaluation of Spatial Autocorrelation and Heterogeneity in the Residuals of Six Regres-sion Models. *For. Sci.* **2009**, *55*, 533–548.
43. Goodchild, M.F. First Law of Geography. *Int. Ency. Hum. Geogr.* **2009**, *81*, 179–182.
44. Zhu, P.; Liu, X.; Zheng, Y.; Wang, S.; Huang, L. Tadeoffs and synergies of ecosystem services in key ecological function zones in northern China. *Acta Ecol. Sin.* **2020**, *40*, 8694–8706.
45. Zhang, Y.; Cai, X.; Yang, C.; Song, X.; Li, Y.; Li, E.; Wang, X. Driving force analysis of landscape pattern changes in Honghu Wetland Nature Reserve in recent 40 years. *J. Lake Sci.* **2019**, *31*, 171–182.
46. Li, J.; Zhou, Z. Landscape pattern and hydrological processes in Yanhe River basin of China. *Acta Geogr. Sin.* **2014**, *69*, 933–944.
47. Zhang, Z.; Chen, S.; Wan, L.; Wan, L.; Cao, J.; Zhang, Q.; Yang, C. The effects of landscape pattern evolution on runoff and sediment based on SWAT model. *Environ. Earth Sci.* **2021**, *80*, 1–12. [CrossRef]
48. Zhang, M.; Wang, J.; Li, S.; Feng, D.; Cao, E. Dynamic changes in landscape pattern in a large-scale opencast coal mine area from 1986 to 2015: A complex network approach. *Catena* **2020**, *194*, 10483. [CrossRef]
49. Qi, D. Study on Evaluation of Forest Ecogical Ecosystem Services of Daxing-anling and Xiaoxing-anling Based on Ecological Location Coefficient. Master's Thesis, Northeast Forestry University, Harbin, China, 2014.
50. Xu, H.; Man, X.; Sheng, H. Water Conservation Function of Different Types of Larix gmelini Forests in Da Hinggan Mountains. *Res. Soil Water Conserv.* **2011**, *18*, 92–96.
51. Diao, Q.; Meng, K.; Zhang, W.; Meng, X. Study on typical artificial forest carbon fixation and oxygen release in Xiao Hing-gan Mountains. *For. Invest. Des.* **2013**, *2*, 26–28.
52. Wang, X. Study on Forest Ecosystem Services of Daxing-anling Forest in Inner Mongolia. Master's Thesis, Inner Mongolia Agricultural University, Hohhot, China, 2014.
53. Liu, L. Study on the Plant Biodiversity and Ecosystem Service Functions Evaluation of Forest Ecosystem In Xiaoxingan Mountains. Ph.D. Thesis, Northeast Forestry University, Harbin, China, 2014.
54. Li, G.; Zhao, W.; Wei, Y.; Fang, X.; Gao, B.; Dai, L. Evaluation on the influence of natural forest protection program on forest ecosystem service function in changbai mountain. *Acta Ecol. Sin.* **2015**, *35*, 984–992.
55. Sui, H.; Song, G.; Zhang, H. Identification of production-living-ecological space at Keshan County level in main grain pro-ducing areas in northern Songnen Plain, China. *Trans. Chin. Soc. Agric. Eng.* **2010**, *36*, 264–271, 323.
56. Yin, L.; Wang, X.; Zhang, K.; Xiao, F.; Cheng, C.; Zhang, X. Trade-offs and synergy between ecosystem services in National Barrier Zone. *Geogr. Res.* **2019**, *38*, 2162–2172.

57. Qi, N.; Zhao, J.; Yang, Y.; Gou, R.; Chen, J.; Zhao, P.; Li, W. Quantifying ecosystem service trade-offs and synergies in Northeast China based on ecosystem service bundles. *Acta Ecol. Sin.* **2020**, *40*, 2827–2837.
58. Han, M.; Dong, X.; Guan, H.; Zhang, Q. Effects of Soil Properties on Ecological Functions in Different Succession Stages of Natural Larix gmelinii Forest in Daxing'an Mountains. *J. Northeast. For. Univ.* **2019**, *47*, 50–54, 89.
59. Yang, S. The Effect of Vegetation Restoration on Ecosystem Services Trade-off and Synergy in the Typical Watershed of the Loess Plateau of China. Ph.D. Thesis, Xi'an University of Technology, Xi'an, China, 2020.
60. Li, L.; Wu, D.; Liu, Y. Hotspots and Trends of the Researches for Land Use and Ecosystem Service—Based on CiteSpace. *Res. Soil Water Consev.* **2020**, *7*, 396–404.

forests

MDPI

Article

REDD+ Conflict: Understanding the Pathways between Forest Projects and Social Conflict

Rowan Alumasa Alusiola [1,2,*], Janpeter Schilling [1,2] and Paul Klär [3]

[1] Peace Academy Rhineland-Palatinate, University of Koblenz-Landau, 76829 Landau, Germany; schilling@uni-landau.de

[2] Institute for Environmental Sciences, Landuse Conflicts, University of Koblenz-Landau, 76829 Landau, Germany

[3] Karlsruhe Institute of Technology, 76131 Karlsruhe, Germany; paul.klaer@student.kit.edu

* Correspondence: alusiola@uni-landau.de; Tel.: +49-06341-280-38554

Abstract: A growing body of literature analyses the conflict implications of REDD+ (Reducing emissions from deforestation and forest degradation in developing countries). However, the way these conflicts unfold is little understood. We address this research gap through the following question: What are the pathways that connect REDD+ projects and conflicts between local communities and other actors? We review 242 scientific articles, selecting eight that allow us to trace how the conflict pathways unfolded. We draw on a political ecology perspective and conceptualize 'conflict pathway' as an interaction of key events and drivers leading to conflict. We find six main conflict drivers: (1) injustices and restrictions over (full) access and control of forest resources; (2) creation of new forest governance structures that change relationships between stakeholders and the forest; (3) exclusion of community members from comprehensive project participation; (4) high project expectations that are not met; (5) changes in land tenure policy due to migrants, and (6) the aggravation of historic land tenure conflicts. Evictions from forests, acts of violence, and lawsuits are among the events contributing to the conflict pathways. To prevent them, the rights, livelihoods, and benefits of local communities need to be placed at the centre of the REDD+ projects.

Keywords: REDD+; conflict; forests; land tenure; political ecology

Citation: Alusiola, R.A.; Schilling, J.; Klär, P. REDD+ Conflict: Understanding the Pathways between Forest Projects and Social Conflict. *Forests* **2021**, *12*, 748. https://doi.org/10.3390/f12060748

Academic Editors: Juan F. Fernández-Manjarrés and Roxane Sansilvestri

Received: 26 April 2021
Accepted: 2 June 2021
Published: 5 June 2021

Publisher's Note: MDPI stays neutral with regard to jurisdictional claims in published maps and institutional affiliations.

1. Introduction

Reducing emissions from deforestation and forest degradation in developing countries (REDD+) is one of the key global instrument for mitigating climate change [1]. Since 2008, over USD 4 billion has been pledged to multilateral climate funds that support REDD+. Approved REDD+ activities have totalled USD 2.4 billion since 2008; in 2018 alone, USD 260 million was approved [2]. More than 50 countries are either in the readiness or implementation phases of REDD+ [3]. REDD+ is not only an instrument that has substantial funding and wide implementation—but it is also ambitious. On the one hand, REDD+ aims to enhance carbon stocks and reduce deforestation; on the other hand, it seeks to benefit local communities [3,4]. However, there is a growing body of literature that questions the positive effects of REDD+ on communities, or even identifies negative externalities, including conflicts within and between communities as well as between them and other actors such as project proponents and government representatives [5–9]. While these studies help to identify the key conflict actors and some conflict drivers, our understanding of how conflicts between local communities on the one hand, and project implementers, companies, and government agencies actually unfold is limited. The present paper addresses this research gap while asking the following key question: What are the pathways that connect REDD+ projects and social conflicts between local communities and other actors?

The literature on the conflict implications of REDD+ can broadly be grouped into two types. The first uses qualitative field research [10–13]. This approach can generate deep insights into specific cases but their applicability to other cases is limited. The second type of research is based on a literature review [14–16]. These reviews can synthesize knowledge and identify general conflict factors but lack detail as to how conflicts actually unfold. We chose an approach that lies between these two types and can combine their strengths. We first reviewed 242 relevant scientific articles, selecting eight that met our criteria for a comparative case study analysis of conflict pathways. A key selection criterion was that the article should provide enough detail on the key events and conflict drivers to enable us to determine how the conflict unfolded. We understand conflict as a situation in which at least two actors perceive their goals, actions, values, needs or priorities as incompatible with each other. In violent conflicts, at least one of the actors uses force to pursue its aim or to directly harm other actors [17].

This combination of literature review and comparative analysis of selected case studies has been successfully applied in a similar study on the role of resource scarcity in local conflicts [18]. Since the distribution of costs and benefits, governance structures and power dynamics play key roles in both REDD+ projects and conflicts, we have embedded our analysis in a political ecology perspective [19].

Section 2 below provides more detail on the theoretical framework, followed by Section 3′s description of the cases and how they were selected. In Sections 4 and 5, we present and discuss the results, concluding how pathways between REDD+ projects and conflict may be avoided and interrupted.

2. Theoretical Framework

We use political ecology as the guiding concept for our analysis because it focuses on how power dynamics influence access to, and control over, natural resources, which are at the centre of the conflicts we aim to understand [20,21]. Political ecology further reminds us that we must pay attention to the governance structures and to the political, historical and cultural embeddedness of dynamics that are often reduced to and portrayed as "resource conflicts" [18]. Finally, the focus on the distribution of costs and benefits across actors and scales is a particularly useful feature of political ecology [22].

Figure 1 shows how we conceptualise the conflict pathway. In the upper part of the arrow, the key events contributing to the conflict pathway are shown in chronological order. In the lower part of the arrow, the key conflict drivers are listed. Conflict drivers are general factors that feed and accelerate the conflict pathway. Or in other words, without the conflict drivers, there would be no conflict. The arrow itself is composed of the interaction between key events and conflict drivers. The text at the end of the arrow indicates whether the REDD+ project aggravated an existing conflict already or created a new one. Using the concept of a pathway to analyse conflicts is not new. For instance, Ide et al. [23] recently used it to disentangle the interactions between water, drought and conflict in the Middle East and North Africa. However, what is innovative, and what we find particularly helpful in retracing how a conflict unfolds is the structure of the pathway figure. It allows us to show, at a glance, specific events as well as general conflict drivers and hence provides a foundation for our comparative analysis of social conflicts related to REDD+ projects.

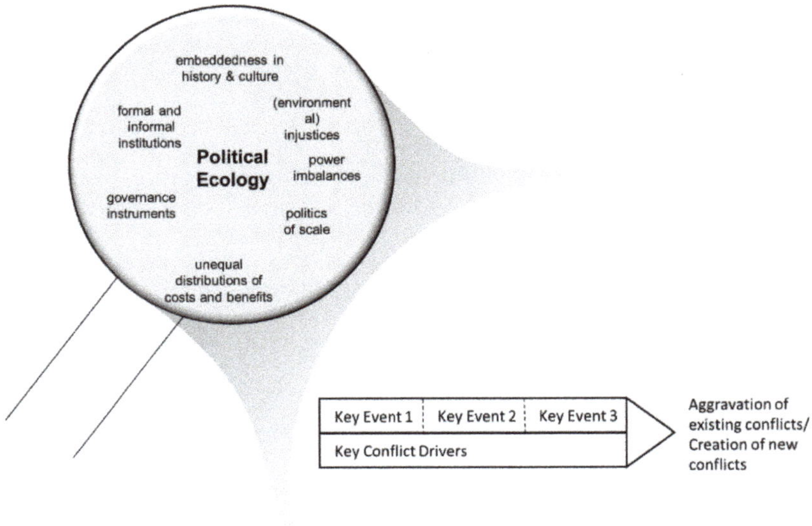

Figure 1. Conceptual framework [19].

3. Materials and Methods

3.1. Selection of Cases

Table 1 shows the selection criteria for the case studies. While the thematic focus is obvious, the level of detail refers to specific information on how the conflict unfolded chronologically as well as a clear identification of key conflict drivers. The selection criteria follow Eisenhardt and Graebner's "replication logic", which implies that " ... multiple cases are discrete experiments that serve as replications, contrasts, and extensions ... " to existing or new theories. The selected case studies enable us to " ... emphasize the rich, real-world context in which the phenomena (in our case social conflicts) occur" ([24], p. 21).

Table 1. Selection criteria for the case studies.

Thematic Focus	REDD+ (Carbon Offset)
Level of detail	High enough to retrace the conflict pathway
Method	Field research
Publication	Peer-reviewed journal article

In November 2020, searching a combination of the topics "REDD*/REDD+/REDD plus" and "conflict*" we found 178 entries in the Web of Science, Scopus, and Google Scholar. To ensure that we do not miss any relevant paper which did not have the word "conflict" in the title, abstract or keywords, we expanded the topic search using the terms "tension*", "harm*", "contestation*", "disappointment*", "disagreement*" and "violence*". This resulted in additional 64 papers. Similarly, we searched for a combination of these terms and "carbon offset*" instead of "REDD*/REDD+/REDD plus", resulting in additional 54 papers.

All available abstracts were read and evaluated by at least two authors. This led to a group of 130 papers that were fully read and again evaluated by at least two authors. Of these, only eight met our selection criteria [25–32]. Seven papers focused on REDD+ projects while one studied a carbon offset project [26].

The level of detail and the research method were the criteria that excluded most articles. The last criterion shown in Table 1 was applied to ensure scientific quality. Originally, we had limited the geographic scope to Africa but, since the number of African cases that matched our criteria was incredibly low, we searched for cases globally. Several of the read papers that did not meet all our criteria were highly useful to strengthen the discussion of the selected case studies [33–36].

3.2. Overview of Cases

The selected case studies are located in East Africa (Ethiopia, Uganda, Tanzania), Southeast Asia (Indonesia, Vietnam) and Panama. Figures 2–4 provide more information on where the case studies can be found in each country. In Vietnam and Panama, only the approximate locations of the case studies are shown in the original papers. This was done to protect the local communities, as the authors of the respective studies explained to us via email. The selection of cases allowed us to analyse the pathways between REDD+ projects and conflicts in a variety of social, cultural, economic, and political environments.

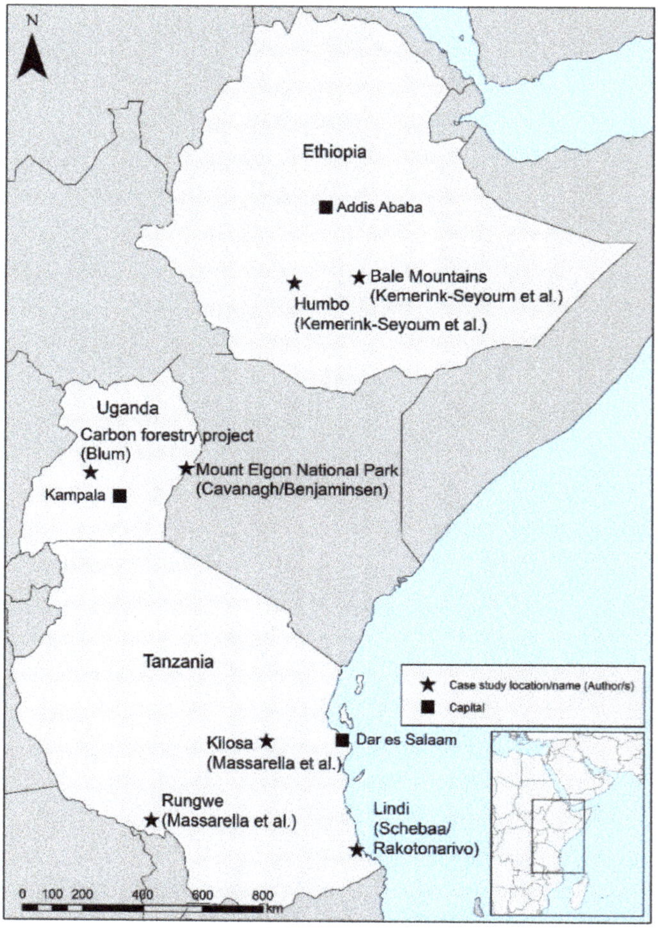

Figure 2. Location of case studies in East Africa (Syed Zulfiqar Ali Shah for the authors, based on [25–27,29,37]).

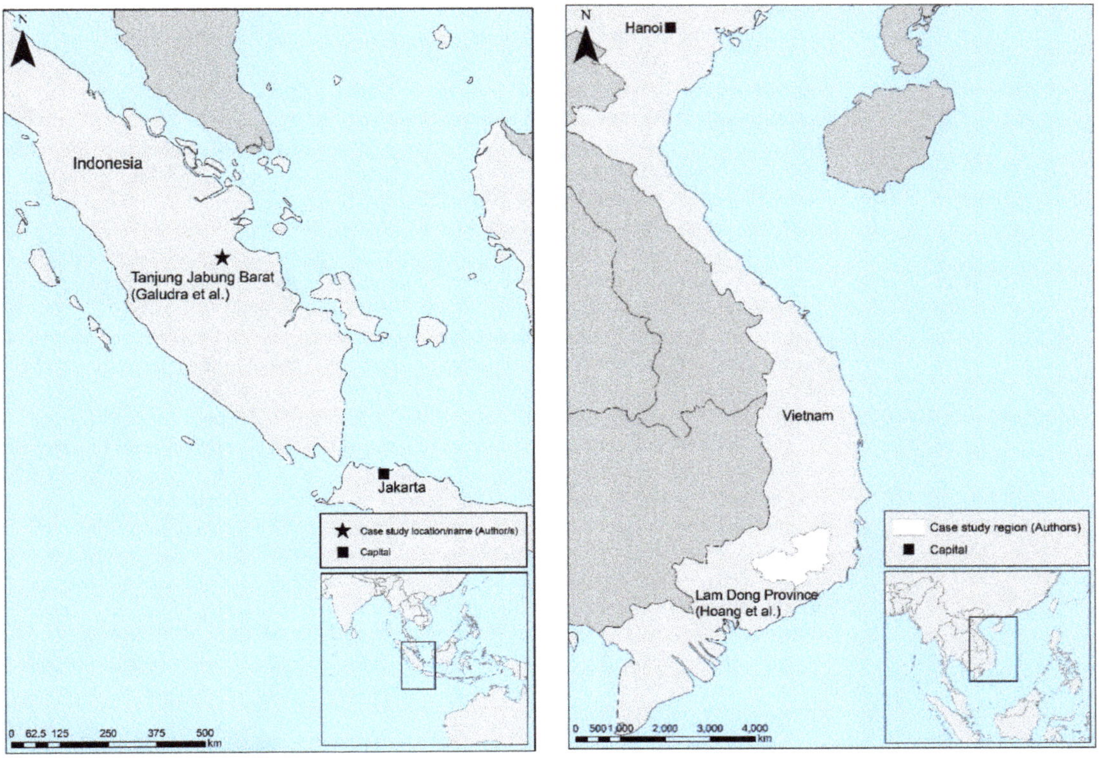

Figure 3. Location of case studies in Southeast Asia (Syed Zulfiqar Ali Shah for the authors, based on [30,32]).

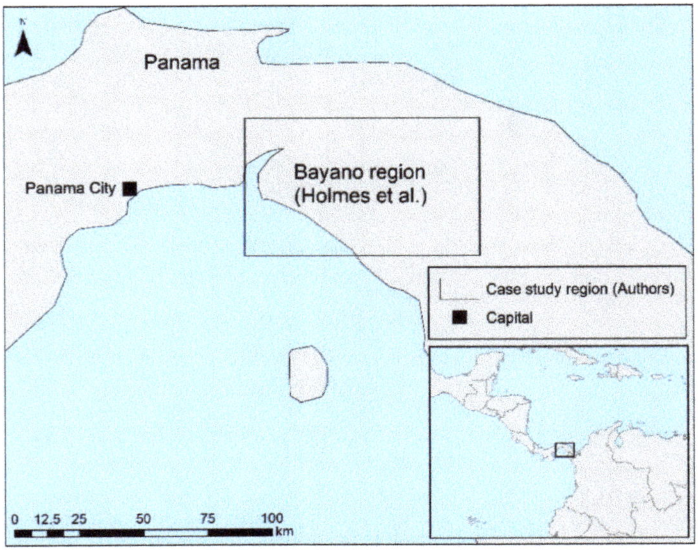

Figure 4. Location of case study in Panama (Syed Zulfiqar Ali Shah for the authors, based on Holmes [31]).

In the following paragraphs, we briefly introduce each case, providing an overview of the REDD+ project as well as the timeline, actors, and intensity of the conflict.

Kemerink-Seyoum et al. [29] investigated Central Ethiopia's reliance on organizational blueprints and how the focus on crafting institutions leads to unequal distributions of natural resources as a result of the enclosure of common land. There was ambiguity in the formation of associations of forest dwellers, leading to ineligible households having to leave the forest. Membership is through a small fee that increased during the project, preventing many people from joining. Non-members were restricted from the forest, leading to divisions between them and the members. The Clean Development Mechanism project involved the reforestation of 2700 ha of degraded communal land which was a source of livelihood to the residents. Some residents, however, opposed the initiation of this project by publicly protesting and marching to the implementing NGO's offices. On a few occasions, these turned into violent clashes with the police, and people being threatened and detained without conviction. The project's negative impacts included reduced fodder supply, which reduced the number of cattle kept by residents. Farmers also had to buy expensive charcoal from a neighbouring village and spend longer hours collecting firewood. They consequently indicated that they felt forced to illegally cut down trees or send their cattle into the forest without permission, leading to fines by the cooperative. Farmers who had been dispossessed of their land took legal steps: the courts ruled in their favour, but the land was never given back. On the other hand, the REDD+ pilot project built on an existing participatory forest management project, with institutional structures which proved to be a major source of tension and conflict. Youth were excluded from association membership and formed gangs that illegally cut down trees and sold timber. This led to violent clashes, causing severe injuries and lasting mental distress, which were not often prosecuted since the gang was feared by most residents. There have been disagreements between the cooperatives and project proponents due to a lack of transparency in benefit sharing and signs of elite capture.

Cavanagh and Benjaminsen [26] focused on a carbon offset project in Eastern Uganda in Mount Elgon National Park and evolving land-use conflicts. In 1992, the Forest Absorbing Carbon-dioxide Emission (FACE) foundation began reforesting degraded sections of the park as a carbon sink. The government opened the market for stored carbon dioxide for FACE to trade as compensation for carbon emissions. At that time there were around 25,000 ha of degraded forest. The process of reforestation at Mount Elgon Park was characterised by the ongoing violent evictions of communities with human right abuses and, in some areas, little or even no prior warning. Paramilitary evictions and controls continued over the next decade between 1993 to 2003. There are no official records of these evictions, but the overall figure is estimated at 6000–150,000 people, who were also not compensated for their loss of land and property or for their injuries. Although communities made accusations against the FACE Foundation and the Ugandan government, FACE denied that its activities had any impacts on land-use conflicts, and the Ugandan government insisted that all the evictions were legal. Mitigation measures were thus not implemented because of these denials of any wrongdoing.

Blum [25] focused on a REDD+ project in Western Uganda. In September 2001, a European forestry company acquired a 50-year license for 12,168 ha of forest reserve from the Ugandan government, to establish a pine and eucalyptus tree plantation as a carbon sink. After the official project start in 2002, conflicts began between the company and the local population who were no longer able to (fully) access the project area to graze animals, cultivate crops, produce charcoal or for cultural purposes. For several years, the company applied a "non-violent but confronting practice" as the CEO of the company phrased it, while "sweeping" ([25], p. 5) the local population out of the project area on a weekly basis. The local population consisted of the members of 20 villages surrounding the plantation and herders migrating to the project area from other parts of Uganda. In 2013, the situation between the company and the local population escalated as company guards burned down a hut located in the project area and beat three locals because they had refused to follow the guards' orders. According to the company, the guards were dismissed, the police investigated the case and after some public pressure was put on the company in 2015, it paid 600 USD as compensation. Nevertheless, sporadic acts of violence still occurred, mostly when herders physically attacked company guards. The study identified limited access of the local population to the forest reserve and plantation as the key conflict driver. In addition, the criminalisation narrative used by the company to characterise the behaviour of the locals who accessed the project area worsened the tensions. Only when the company—partly as a response to criticism by the Forest Stewardship Council (FSC)—chose a more cooperative path did the relationship with the local communities improve.

Massarella et al. [27] focused on two REDD+ pilot projects in Tanzania (Kilosa and Rungwe villages) to analyse the expectations of the people involved in pilot projects that had different approaches to implementation. These projects were active between 2009 and 2014. At the national level, there were high expectations: strong statements were used to show the benefits of REDD+ to the people but the pilot projects began before key actors were fully aware of what REDD+ involved. Strong positive and negative expectations existed at the village level: some villagers feared eviction from their land which would be taken by Europeans, while others feared that wild animals would be introduced in their area. Village Land Forest Reserves (VLRFs) were then established and gazetted. VLFRs committees and leaders in both villages then evicted people with farms in the reserve areas. Kilosa's elevated expectations were influenced by the preliminary stages of project activity, with villagers receiving their first trial payment, the building of the project office and the establishment of some livelihood activities. However, due to the relocations that affected many people, conflict erupted between village leaders and people refusing to be evicted. The villagers were split between those who were for and those against the move. The conflicts continued even after the end of the project: about 25 farmers continued to farm in the VLFRs; threats of violence were reported by both parties and farmers who refused to move were taken to court. In Rungwe, however, the implementing NGO decided not to focus on the REDD+ hype. The NGO had concerns about " ... making promises to communities that you can't deliver" ([27], p. 380). Awareness of and participation in the REDD+ pilot project was only extended to village governments and committees. The village level actors had low expectation from the project. The number of people evicted from the VLFRs was also incredibly low and thus conflict was not experienced. The comparison of the experiences of the two villages emphasises the need for the recognition of trade-offs and their potential consequences for villagers.

Scheba and Rokotonarivo [37] investigated why REDD+ initiatives result in social conflict over land-use and negative outcomes for some stakeholders in south-eastern Tanzania, despite project proponents' commitments to pro-poor outcomes, social safeguards and good governance. Contestation over land boundaries and forest are frequent in this area. Forest loss was at 1.9% per annum in 2000 and 2006. Livestock keeping and small-scale agriculture through clearance of forestland are the main livelihood activities. In 2009, a REDD+ project commenced to reduce more than 110,000 t of carbon dioxide emissions. An emphasis during project implementation was put on obtaining free, prior and informed consent of project communities. Despite all these efforts, conflicts between villagers and local government emerged, due to unclear boundaries. The local government then threatened to put up a new land boundary by force. Much consultation and many project proponents raised the expectations of villagers concerning income from REDD+, so an agreement was reached. Forest bylaws were promulgated, governing the extraction of both wood and non-wood forest resources. New structures for forest governance at the village level were established. An equal payment scheme was adopted by the project to make trial payments, with the assumption that everyone should contribute and benefit equally from forest conservation. After this, however, inter-community conflict and contest restarted over forest ownership. Further confrontations arose when one village chased away farmers from another village from farming on their forest land; eventually, police dispersed them. Several conflicts were reported between other villages over land boundaries due to the benefits of REDD+. Eventually, the project proponent, the local government and community members worked together to establish legitimate land boundaries. One farmer was, however, notably distressed over his lack of access to his farm, and many attempts to obtain justice from the village forest committee failed.

Hoang et al. [32] focused on a REDD+ pilot project in Vietnam's central highlands to explore the dynamics of conflict over forests. The area is inhabited by migrants and original landowners. Before 1975, the forest was customarily managed by three K'ho clan heads and the symbolic owners of the land. The community freely used the forest for cultivation, timber extraction, gathering forest products, and hunting. The state then abolished traditional customary rights to land and forest, and the K'hos' fallow land was considered "ownerless" ([32], p. 26). A state-sponsored migration programme then relocated thousands of people to the central highlands. They were allowed into the protected forest area for small-scale logging, cutting down small trees for house construction, and collecting firewood. Between 2003 and 2008, the forest sector changed, with a "red book" ([32], p. 26) (land certificate) being required for families to have rights over forest land. A Payment for Environmental Services (PES) programme resulted in the establishment of new and stricter rules for forest management and protection. Disputes erupted between the villagers and patrols because the former continued to exploit timber. Villagers and people from the surrounding areas "illegally" ([32], p. 27) encroached on the forest to grow coffee. The implementation of the REDD+ project was split into two phases—2009 to 2012 and 2013 to 2018—and was built on pre-existing institutional PES structures. REDD+ activities focused on about 55% of village households and commune officers. Some of the K'ho involved in PES raised the issue of injustice at different forums but their concerns received no response. Debates ensued over the agricultural development loan because the poorest households never met the criteria. Land tenure was also discussed regarding forest resources and PES. The K'ho regarded REDD+ as ineffective and oblivious to local norms of social equity. In interviews, they said, "REDD+ cares about trees, not about people" ([32], p. 30). They also engaged in "everyday forms of resistance" ([32], p. 30). They illegally cut down trees from the forest in the middle of the night or when the officials were on a break. Families with coffee plantations bordering the forest would expand their land by cutting trees each year around the forest boundary to increase their land.

Galudra et al. [30] showed how migrant communities to peatland in Jambi, Indonesia changed the dynamics of relations, land tenure and forest conservation. TanJaBar area changed land tenure from the 1970s to the present. The land was initially customarily owned but is currently both communally, privately and government-owned. The tenure changes are due to the influx of migrant communities; these migrants also changed the activities on the peatland from the original peatland forest. Conflicts between the local community and government started in 1997 when the Ministry agreed to increase the concession in the TanJaBar area from 35,580 ha to 43,750 ha. The area claimed was categorised as a non-forest area from 1993, although the government said the area is conversion production forest. In 2002, the land was demarcated, and 7224 ha of community land was identified as part of the new production forest area. The area belonged to migrants, some of whom had received documentary proof of ownership. The community held protests, but the company proceeded to convert the area into an acacia plantation. Conflicts became violent in 2010 at the "hot spot" ([30], p. 719) village of Senyerang. When about 1500 people demonstrated, two men were shot and one of them died. With political elections in 2010, one candidate promised the villagers that their land would be given back. After the candidates' victory, the community members became more aggressive. The land was then converted back to non-forest through the enactment of a new land-use plan. The company was reluctant to undertake any negotiations but bowed to pressure and the villagers became co-managers of the rubber plantation. While the villagers gained rights and access to the land, details of benefit-sharing remained under negotiation. In another village, conflict over access to forest peatland was reported. The regency forest agency (RFA) tried to claim the land back, but the migrants resisted the programme and it failed, prompting

RFA to consider changing the status of the land to non-forest. This will, however, lead to conflict with the national government due to their national emission reduction objective. The author emphasizes that successful REDD+ initiatives require a clear and secure tenure and that migrants influence over land tenure and relations with other stakeholders ought to be recognised in REDD+ projects.

Holmes et al. [31] show how a small bottom-up forest carbon-offset project which later became a 19 ha REDD+ project was impacted by existing land conflicts between members of an indigenous community and migrant farmers of Latino origin, called "colonos" ([31], p. 4), in Emberá, Eastern Panama. The study identified monetary compensation as the main interest of the community for participating in the REDD+ project. The conflict between the two groups began in the mid-1970s when the government displaced 400 indigenous Emberá and 2500 colonos to develop a hydroelectric dam. Both communities needed new land but, later, only the Emberá were officially given access to land while the colonos were not, losing their farms and grazing land. Tensions between the two groups heightened in 2009 when signs were put up, and a border and a community-based patrol were established to demarcate and protect the area for the REDD+ project. The colonos responded with threats of violence toward the Emberá and they continued to clear the forest inside the demarcated area. The conflict was resolved through the formation of an Advisory Council on Conflict Resolution that included local and national stakeholders. The council developed seven recommendations that mainly focused on the key underlying conflict driver: unclear land rights and confusion " ... about the roles and responsibilities of government agencies in land law enforcement ... " ([31], p. 10).

4. Results

We first describe the conflict pathways (Section 4.1) before presenting the effects of conflicts on local communities (Section 4.2) and their responses (Section 4.3), as well as conflict mitigation measures (Section 4.4).

4.1. *Conflict Pathways*

Key drivers of the identified conflict pathways include (1) injustices and restrictions over (full) access to and control of forest resources; (2) creation of new forest governance structures that change relationships between stakeholders and the forest; (3) exclusion of community members from comprehensive project participation; (4) high project expectations not being met; (5) changes in land tenure policy due to migrants; (6) aggravation of historic land tenure conflicts (Figure 5). There are differences, similarities and overlaps in the pathways.

Injustices and restrictions over access and control of forest resources were key conflict drivers that fed the conflict pathways in all case studies. Local community members were prohibited from accessing the project area for either farming, grazing animals, firewood, timber, or dwelling. However, one difference was noted in the Ethiopian case study. There, the membership fee for the cooperative societies was increased to reduce the number of those who could participate in the project and thus access forest: only association members were given (partial) access. Community members reacted differently to the restrictions that led to conflict (see Section 4.3). Elite capture was noted in influential members with connections to those in leadership who were allowed to (fully) access the forest [29]. The creation of new governance structures changed the relationship between stakeholders and the forest and led to the introduction of new powerful stakeholders—for example, the village council, the local village natural resource committee and the village assembly. Discrimination and injustice towards some villagers by the new village leadership were recurrent. Many people who sought justice for the loss of land and property from the leadership were unsuccessful, causing some to resort to the courts [27,37].

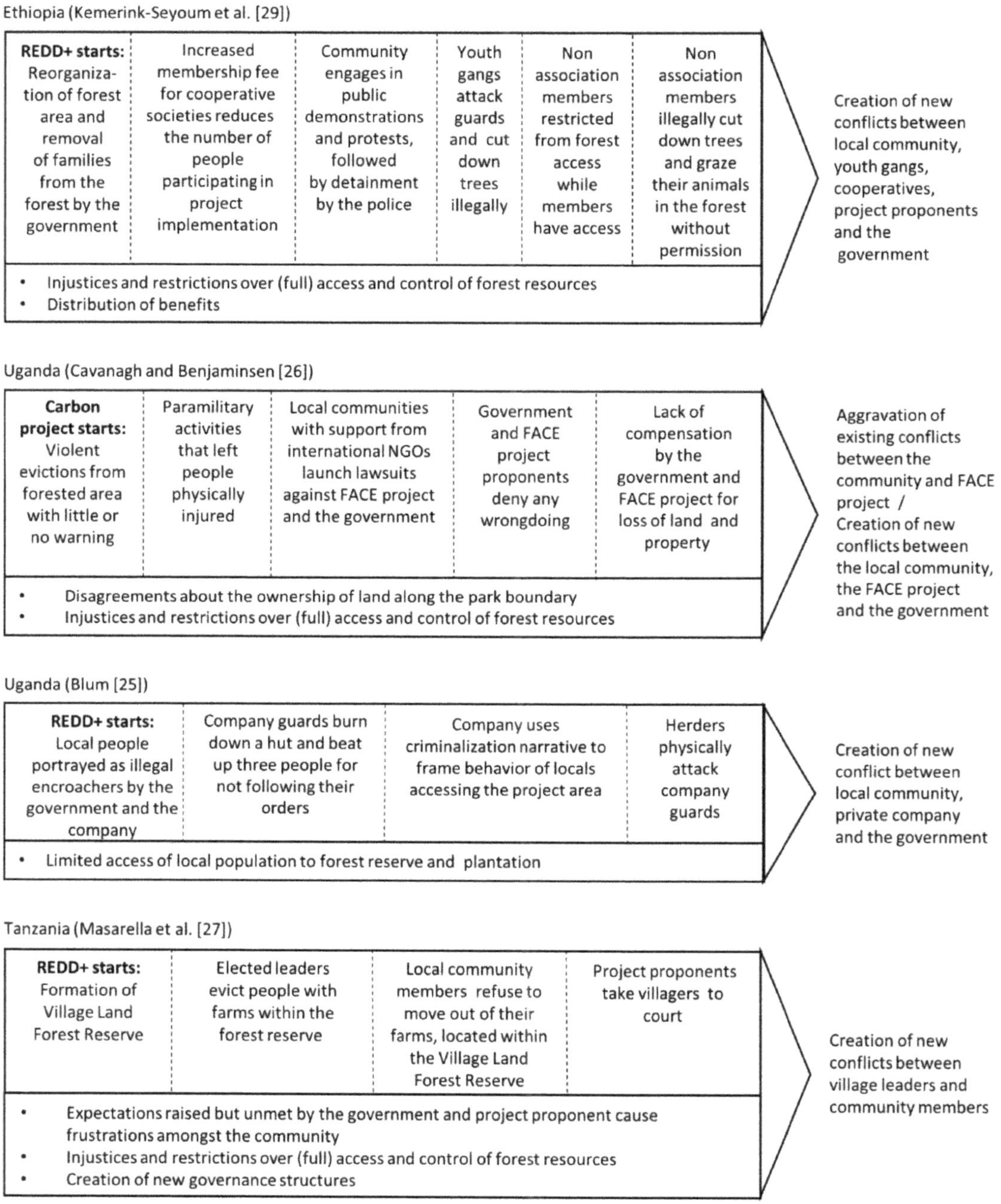

Figure 5. *Cont.*

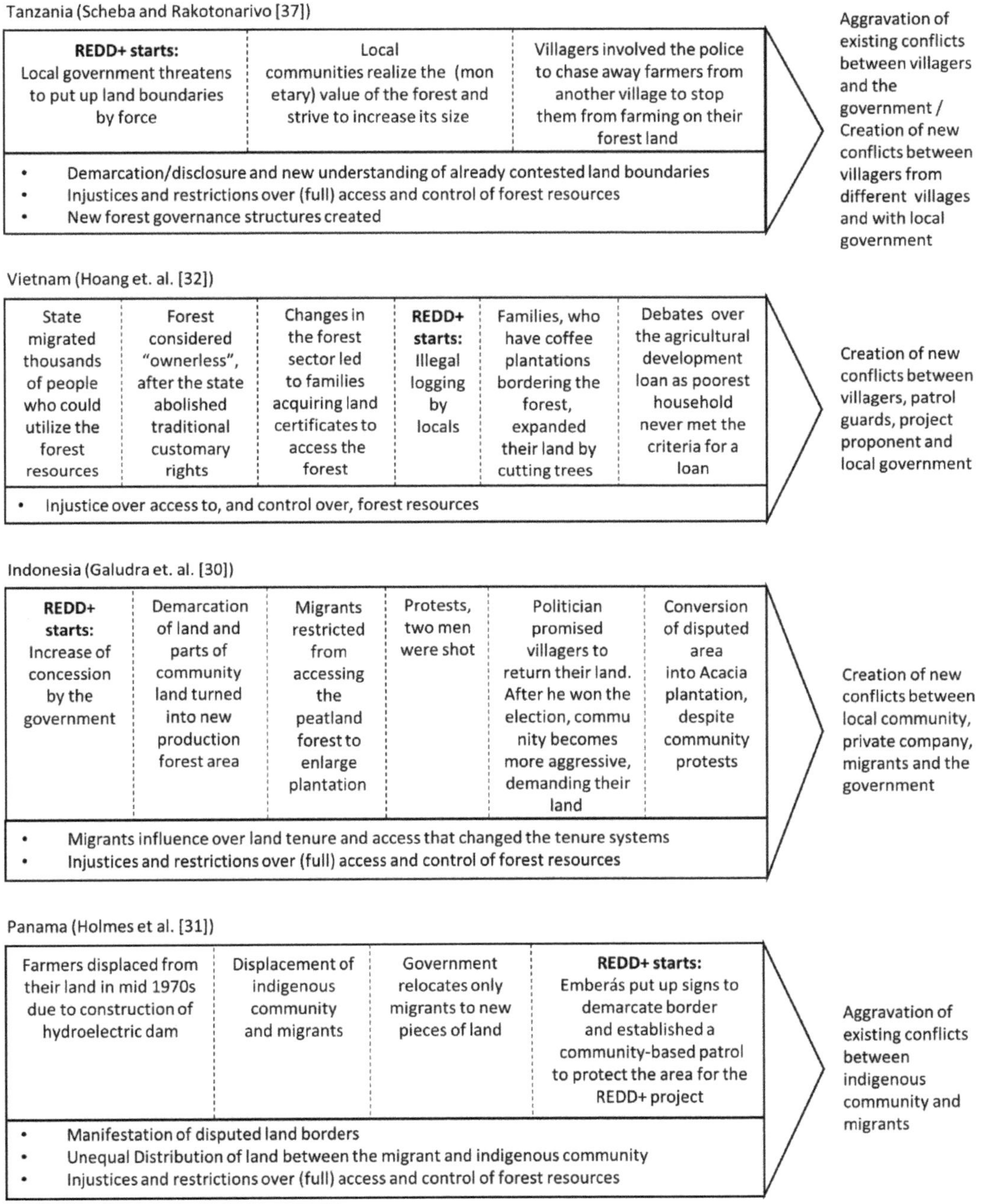

Figure 5. Conflict pathways (the authors).

Land tenure fed the conflict pathway in four cases. Two dynamics were identified: (i) migrants' influence over land tenure changed the interaction between stakeholders; (ii) the re-emergence of existing land tenure conflicts due to the implementation of REDD+ projects. Migrants were seen as intruders by indigenous communities because their presence changed the land tenure system and land utilisation, leading to conflict [30,31]. In Indonesia, migrants searched for more land despite being allocated holdings by the government while, in Panama, migrants were relocated after being displaced while the indigenous community was not. Scheba and Rokotonarivo [37] describe how a REDD+ project increased understanding of the importance of forests and their boundaries. Community members had a strong urge to conserve bigger pieces of forest with the expectation of higher carbon income. This resulted in conflicts with other villages. Historically, the four case studies experienced changes in political leadership that altered policies and legislation on land tenure systems. This led to misunderstanding and a lack of clarity for the community members as to who owned the land, creating tension and conflict. Confusion also existed regarding the roles and responsibilities of different government agencies in land law enforcement, derailing conflict resolution processes. For example, Hoang et al. [32] described the inability of two major agencies to resolve conflict because they lacked clarity on their mandate to address conflicts over land and land use.

Community members' participation in the REDD+ project was based on their hopes for high returns. Kemerink-Seyoum et al. [29] and Massarella et al. [27] described how national governments and project proponents invested heavily in raising local people's expectations on the project's benefits—which were not eventually met. These two case studies, however, used different approaches to raising community expectations. In Tanzania, local campaigns were conducted through the media and through roadshows to advertise the benefits of the REDD+ project to be implemented in a few selected sites. Unmet expectations then caused anger, a sense of betrayal and undermined the support for the conservation projects [27]. Similarly, in Ethiopia, cooperative members had been promised cash payments based on their shares. However, only well-connected members in some cooperative societies received financial gain through carbon benefits, leading to discrimination and anger amongst members [29].

The differences in the conflict pathways reflect the diverse perspectives of the forest-dependent communities towards land tenure. For example, four case studies had undergone changes in land tenure from communal to private ownership and then to government-owned. Before the implementation of REDD+ projects, forests in Tanzania and Indonesia were regarded as open access since the land was communally owned. The use of forest resources was viewed differently due to the different activities that communities practised. In all case studies, the forest was accessed for subsistence use. In two case studies [25,29], cattle herding communities depended heavily on the forest for grazing their animals; in three cases [30,32,37], communities owned plantations for palm oil, coconut, coffee and rice within the forest. Relationships and interactions with stakeholders from outside their territory shaped communities' reactions to the project. In one unsuccessful carbon offset project in Uganda [26], the local community experienced eviction by National Resistance Army paramilitaries, creating a sour relationship between communities, project proponents and the government. In seven studies, evictions were conducted by newly elected local leaders who sometimes involved the police or project proponents, creating division between those who were for and against the project. Lack of in-depth project understanding was also an issue in all the case studies. The local communities had an understanding of how the REDD+ project would positively change their livelihood while other important project features—for example, carbon markets and how prices were set—were not understood by the communities. This created agitation due to unmet expectations. Further differences were due to the diversified strategies used by the project proponents for project awareness, planning and implementation, and the varied stakeholders involved in project implementation. In one case study, an extensive community engagement strategy was used but conflicts occurred due to land tenure security, while, in seven of the case studies, community members were partially involved in the project cycle. This led to conflict in one case study, while six experienced conflicts due either to restricted forest access or new governance structures at the local level. Interestingly, Rugwe village in Tanzania did not experience conflict because very few people were evicted from the project area and the project proponent did not inform the community about the benefits of REDD+. [27]. The major conflict actors included indigenous communities, migrants, government departments at the national and local level, and the project proponent (Figure 5). In some cases, cooperative societies, youth gangs and patrol guards were also involved.

In summary, we observed that the occurrence of a conflict caused by all the stakeholders involved in a project contributed to accelerating conflict pathways. Project proponents and governments contributed to the conflict by restricting communities' access to forests and full project engagement. Community reactions to these restrictive measures eventually led to conflict. In addition, the current and potential impacts of REDD+ on local communities often disrupted people's livelihood strategies, institutions, and socio-cultural systems, eventually leading to conflict.

4.2. Effects of Conflict on Local Communities

The analysis of the eight case studies shows that most of the local communities were severely affected by the conflicts, with some experiencing fewer negative impacts on their livelihood. More violent conflicts had more severe effects on local communities. Furthermore, it is evident that forest resources are a source of livelihood to all the communities in the eight case studies. They use these for farming, grazing animals, gathering firewood, timber, and non-timber forest products (NTFPs). We are in full agreement with other authors [38–40] who acknowledge that poor and indigenous people are vulnerable as they are often highly dependent on forest resources for their livelihood. However, due to conflict, local people can no longer (fully) access these resources. The effects of conflicts on local communities and their responses are summarised in Table 2.

In five out of the eight case studies, people lost farming land which was previously in government, private or communal ownership. The land in one project area was used for commercial oil palm farming. In four of the cases, the community used the land for subsistence farming. New restrictive rules and regulations on forest-related livelihood options led to a decline in forest-based income and reduced yields, leading to food insecurity and poor nutrition. Scheba and Rokotonarivo [37] give an example of a farmer who was distressed over his lack of access to his farm which was his sole source of livelihood. Many attempts to get justice from the village forest committee did not succeed, leaving him more vulnerable than before the project's implementation. Additionally, in seven case studies, the community was left with either physical or mental injuries, with some living in fear of being attacked by youth gangs and others being taken to court by the local government. In one case study, the conflict led to two people being shot, one died. Conflicts between different stakeholders led to divisions between those supporting and opposing the project. In four cases, the division was seen either within communities–between lower and higher social class households, the project proponents and other community members–or between the local community and the government. These divisions have led to further mistrust and injustice towards those opposing the project. For example, in Ethiopia, people who were against the project were denied positions in cooperative societies that ran the project. Additionally, in three of the case studies, communities experienced the loss of grazing land. In Ethiopia, due to the lack of grazing land, the prices for fodder drastically rose and farmers decided to sell some of their animals. Loss of land and property was a major issue experienced in four of the case studies. In two different projects from Uganda, people were violently displaced from their homes within the project area and their houses torched. Lastly, firewood and timber for subsistence use were not easily accessible. In Ethiopia, it was reported that the prices for firewood rose steeply. In Vietnam, the villagers illegally stole timber to build their houses.

Table 2. Effects of conflicts on local communities and their responses.

Country (Authors)	Effects of Conflict on Local Communities	Communities' Response
Ethiopia (Kemerink-Seyoum et al. [29])	Loss of • land/physical displacement. Division between • association members and non-members. Other effects • Immigrants evacuated from the forest back to their original homes. • Some community members are living in fear of being attacked by a youth gang. • Decline in the supply of fodder for animals, leading to increase in its price. • Decline in daily household nutrition and resilience for dealing with negative events. • Increase in price of firewood. • Severe physical injury and mental stress. • Detention without conviction by court of law. • Court decision for farmers to be compensated with their land not honoured. • Threatened with exclusion from access to public services. • Relinquishing positions held in the association.	• Farmers reduced the number of cattle they own. • Youth excluded from membership formed gangs that illegally cut down trees in the protected forest area. • When caught, gang members attacked forest guards with knives. • Illegal loggers bribed police to carry logs without licenses. • Gang members and project proponents taken to court by the community. • Fear of testifying against wealthy members of the association due to their influence on the association's decisions. • Walking longer distances to the neighbouring village to collect firewood.
Uganda (Cavanagh and Benjaminsen [26])	Loss of • land for crop cultivation. • property through evictions. Other effects • Injury and physical pain from paramilitary evictions. • Basic rights to access common property resources denied.	• Litigation against the government for seizing ancestral land. • Encroachment into the protected area.
Uganda (Blum [25])	Loss of • property: houses burned down. • grazing fields. • livelihood/increased poverty. Other effects • Physical pain caused by beating received from forest guards. • Defiant people taken to the police station to pay large fines while their cattle were confined and often injured.	• Migrant pastoralists and locals' resort to resistance and violence to access the restricted area. • Disobeying authority by continually "encroaching" ([25], p. 6) on the restricted areas. • Hunters and herders sometimes start fires in restricted areas to encourage the growth of new grass for their animals. • Farmer helped other cattle herders from 20 km away enter the forest reserve. • Afraid to approach project proponent. • Refuse dialogue with project proponent when they are wrong.

Table 2. *Cont.*

Country (Authors)	Effects of Conflict on Local Communities	Communities' Response
Tanzania (Masarella et al. [27])	Loss of • source of livelihood. • farmland through evictions. Division • within villages. • across villages. Other effects • Defiant farmers taken to court.	• People refuse to move from their land. • 25 farmers defiantly continue to farm in restricted areas. • Some villagers feel disappointed, cheated; lose confidence in the pilot project and any other future project. • Blaming village leaders as a source of the project's failure.
Tanzania (Scheba and Rakotonarivo [37])	• Anger, discomfort, and threats of economic and physical displacement. • Restricted access to farms.	• Villagers fight to increase their forest size and thus potential carbon income. • Villagers seek assistance from the district or police in their attempts to displace "illegal squatter" farmers. • Adversely affected farmers individually contest majority consent to expand forest reserve area.
Vietnam (Hoang et. al. [32])	Loss of • farmland.	• Illegally felling trees in the forest. • Illegally expanding farming land at the forest edge.
Indonesia (Galudra et. al. [30])	Loss of • land. • livelihood from oil palm farming. • life: two people shot, one fatally.	• Increased aggression in retrieving lost land. • Opposition to new forest conservation programmes from the government. • Demonstrating against the pulp and paper industry.
Panama (Holmes et al. [31])	Loss of • farm and grazing land to the migrant community. Division between • indigenous community and migrants. Other effects • Farmers paid a fine of \$1000–\$1500 and ordered to cease any "environmentally detrimental" activities.	• To prevent migrants from accessing the project site, forest landowners: (i) posted signs to delineate deforestation parcels to avoid; (ii) trained a community-based patrol to ensure compliance; (iii) established a reforestation border in the conflict area. • Migrant community continued to clear an estimated 36 ha. of forest. • Community leaders initiated dialogue to resolve conflict.

4.3. Community Responses to Conflict

Community responses to conflict differed according to the land tenure system (communal, government or privately-owned land), whether the conflict was violent or not, past experiences with conservation projects, and already-existing conflicts before project implementation. These responses were either by individuals or groups. Some private landowners and community leaders developed strategies to resolve conflict, though the process failed and the conflicts continued [31]. On the other hand, in five case studies, communities on communally-owned and government land reacted more aggressively against government officials, project proponents and the immigrants [25,26,29,30,37]. In five case studies, community members took part in demonstrations, attacked forest guards and, in some instances, beat them up or threatened them with knives, organised for migrant herders to invade the protected areas, or started fires in the project area, while others went to court. Youth gangs were also formed in one case study: they involved themselves in illegal logging and were seen as a threat to the entire community. However, in three case studies, the communities decided to engage in "power to resist" ([25], p. 2) by illegally felling trees, farming on either the edges or within the protected area, and refusing to participate in new projects or dialogue as suggested by the government or project proponents [27,31,32]. In one turn of events, the Vietnamese local state forest management agency was involved in illegal tree felling from the protected project area due to conflicts that had arisen between them and the project proponent; this demonstrates the power held by the local government vis-à-vis global intervention [32]. In three project sites, local communities were both actively aggressive and engaged in everyday resistance [25,29,30]. Furthermore, illegal loggers without licenses bribed the police, some people walked for longer distances to collect firewood, and some community members, due to fear, chose to remain silent rather than testify against the government, wealthier groups, or youth gangs. Finally, a feeling of disappointment, of being cheated and a lack of confidence in the conservation programmes were observed in two case studies [27,30].

4.4. Conflict Mitigation Measures

Conflict mitigation measures were implemented in six of the eight case studies. The measures were implemented either by project proponents in collaboration with local and national governments and local communities or by a court ruling. It was observed that in five case studies, mitigation measures were implemented because of litigation against the government and project proponents, criticism by the media and human rights organisations, reviews and recommendations by international auditors, and external pressure from other stakeholders [25,26,29–31].

Two case studies had courts of law rule in favour of the community members, who were then allowed to return to their farms, with further evictions and destruction of property ceasing [26,29]. In cases of land tenure conflict, the project proponents worked together with the government and the local community to develop legitimate land boundaries and integrated land-use practices with the project [30,37]. Community members were also fully involved in the project cycle and in co-managing a private rubber plantation company operating within the project area. Furthermore, access to restricted forest areas was regulated through the formulation of a formal protocol. Community members were permitted partial access to the project area for their livelihood. The formation of an advisory council on conflict resolution and an emphasis on peaceful co-existence between "good neighbours" [25] was achieved by project proponents and the government [25,31]. In the case study that featured conflict over the sharing of benefits, the cooperative made the unilateral decision to pay cash to their members under the pretext of micro-credit services, an equal payment scheme created, and benefit-sharing of the rubber plantation was under negotiation [29,30,37]. Finally, in Uganda, the project proponent built boreholes for the community to freely access water for their animals and daily use [25]. In two case studies, no mitigation measures were implemented [27,32]. Table 3 provides a summary.

Table 3. Conflict mitigation measures.

Country (Authors)	Conflict Mitigation Measures
Ethiopia (Kemerink-Seyoum et al. [29])	Court ruled in favour of farmers as legal tenants of land, and they should be allowed to return to it. Forest cooperative's unilateral decision to pay cash to its members under the pretext of micro-credit services.
Uganda (Cavanagh and Benjaminsen [26])	Project subjected to a series of independent examinations by one of the world's largest and most respected audit firms. Favourable consent judgment that recognised the community as the "historical and indigenous" inhabitants of the Mount Elgon Forest. Further evictions and destruction of property stopped by the court.
Uganda (Blum [25])	Regular meetings between government and local community to clarify land titles. Changes in the enforcement policy from a strict one-to-one dialogue. Collaborative livelihood support programmes established. Dismissal/reduction of the number of forest guards. Grievance procedure to investigate conflict claims was carried out by a carbon market consultant. Company stopped tree planting on 1000 ha. of land, contested by neighbouring farmers. Matter forwarded to NFA to seek clarity about its owners. Depending on the age of planted trees, the company tolerates cattle grazing within the reserve. Company built six boreholes and 15 valley dams close to the project area for the farmers to freely access water. Access to restricted area regulated by formal protocol. Collaborative monitoring with local leaders. Company emphasizes peaceful co-existence between "good neighbours", as well as attracting enough rain for the future.
Tanzania (Masarella et al. [27])	No mitigation measures implemented,
Tanzania (Scheba and Rakotonarivo [37])	New forest governance developed and implemented. Equal payment scheme created. Legitimate land boundaries developed. Integrated land-use planning exercises discussed, mapped, and documented.
Vietnam (Hoang et. al. [32])	No mitigation measures noted.
Indonesia (Galudra et. al. [30])	New land-use plan developed. Villagers co-manage a private rubber plantation company. Benefit-sharing of rubber plantation under negotiation.
Panama (Holmes et al. [31])	Formation of Advisory Council on Conflict Resolution and REDD+.

5. Discussion

This paper takes a political ecology perspective to analyse the conflict pathways of eight REDD+ projects in sub-Saharan Africa, Southeast Asia, and Panama. Six key conflict drivers which have accelerated these conflict pathways were identified: (1) injustices and restrictions over (full) access and control of forest resources; (2) creation of new forest governance structures that alter the relationships between stakeholders and the forest; (3) the exclusion of community members from comprehensive project participation; (4) high project expectations which are not met; (5) changes in land tenure policy (party due to migrants); (6) aggravation of historic land tenure conflicts. Several events caused by either the government, project proponents or the local communities further contributed to the conflict pathways. Forest-dependent community members were restricted from forest access because they were labelled "illegal encroachers" ([25], p. 5). This led them to embrace "power to resist" ([25], p. 2.) through engaging in public demonstrations, assaulting forest guards, organising migrant herders to graze their animals in the forest reserve, and illegally encroaching the forest reserve for farming and logging. In most instances, they were not successful since they were opposed by elected local and association leaders, the government and project proponents who had more power. Forest conservation projects change relations of power, accountability and representation between stakeholders [41,42]. Our findings are therefore in line with other authors in illustrating that power disparities within local communities lead to inequitable benefit-sharing across actors [43–46]. REDD+ ought not to forcefully remove the rights of existing communities nor restrain local forest access [47]. To ensure equitable development and socio-environmental justice, communities should be equal stakeholders with full information access and consent to REDD+ [35,48]. Migrants have influence over land tenure, changing the policies and interaction between stakeholders and exacerbating existing land tenure conflicts. Land tenure security is a major challenge for REDD+ projects [49–51]. Our findings support scholars who argue that rights over land and forests are layered and overlapping, often contested and usually ambiguous, and exist in a pluri-legal setting [52–54]. Tenure arrangements play a critical role in determining equitable participation in and access to REDD+ benefits [55–57] and the attainment of social and livelihood outcomes [50,58–60]. The clarity and security of land tenure are key to the success of REDD+ and access to its benefits. We identified signs of forest decentralisation in some case studies, in line with the concerns raised by several authors [61–65]. The creation of new governance structures empowers and authorises those in leadership and hence creates a potential for abuse.

To prevent abuse of power and the described negative effects of REDD+ projects on local communities, the United Nations Framework Convention on Climate Change (UNFCCC) has guidelines and safeguards in place. One of the safeguards demands "Respect for the knowledge and rights of indigenous peoples and members of local communities, by taking into account relevant international obligations, national circumstances and laws, and noting that the United Nations General Assembly has adopted the United Nations Declaration on the Rights of Indigenous Peoples" ([66], p. 26). Further, the UNFCCC calls for "The full and effective participation of relevant stakeholders, in particular, indigenous peoples and local communities [...]" ([66], p. 26). Contrasting these demands with the findings of our study, one can conclude that the safeguards in the analysed case studies were ineffective because they were either disregarded or poorly monitored and implemented. Approaches such as Free Prior and Informed Consent (FPIC) had very limited success in empowering local communities as examples from the extractive industry have shown [19]. One reason is that FPIC is often seen by project proponents and implementers as a way to legitimatise a project and to minimise risks, rather than to actually protect indigenous or community rights [67]. Milne and Mahanty ([34], p. 133) provide detailed insights into the techno-bureaucratic regime behind REDD+ and its "audit culture" which they call "apolitical and indifferent to local realities". Based on their case study on a REDD+ project in Eastern Cambodia, Milne and Mahanty ([34], p. 133) even argue, that the certification and monitoring mechanisms behind REDD+ allowed the national government to exert "bureaucratic violence" on local communities to silence criticism and implement unjust (project) measures. Similarly, Pasgaard ([35], p. 122) identifies a "risk of procedural inequity in REDD+". Inequity and inequality also determine whose values matter [36]. In extreme cases, this can lead to a situation in which local community members "[...] see themselves as subjects of controlled lab experiments, or guinea pigs, rather than as having agency and control of their options [. . .]" as Sanders ([33], p. 79) conclude based on their case study on a REDD+ pilot project in Indonesia.

Against this background, we not only call for an assessment and revision of the REDD+ safeguards and their implementation but more generally for a shift in how local communities are considered in REDD+ projects. Rather than seeing or portraying locals as obstacles who, at best, need to have some co-benefits and "effectively participate", it is local communities and their rights, livelihoods and benefits which need to be placed at the centre of each REDD+ project. This implies switching the priorities of REDD+, which should thus become community development programmes with some carbon reduction effects rather than the other way around. Without this change, REDD+ will continue to undermine local livelihoods and (re-)produce conflict pathways.

Author Contributions: Conceptualisation and methodology, R.A.A. and J.S.; formal analysis, R.A.A., J.S. and P.K.; writing—original draft preparation, R.A.A., writing—review and editing, J.S. and P.K.; funding acquisition, R.A.A. and J.S., supervision, J.S. All authors have read and agreed to the published version of the manuscript.

Funding: The research for this paper was funded by an Alexander von Humboldt Climate Protection Fellowship and by the University of Koblenz-Landau. The APC was covered through the Open Access Fund of the University of Koblenz-Landau.

Institutional Review Board Statement: Not applicable.

Informed Consent Statement: Not applicable.

Data Availability Statement: The data presented in this study are available on request from the corresponding author.

Acknowledgments: We would like to thank the external reviewers and Anne Hennings for their helpful comments and suggestions. Further, we thank the authors of the case studies we contacted for responding to our questions. We thank Zulfiqar Ali Shah for creating the maps.

Conflicts of Interest: The authors declare no conflict of interest.

References

1. Asiyanbi, A.; Lund, J. Policy persistence: REDD+ between stabilization and contestation. *J. Political Ecol.* **2020**, *27*, 378–400. [CrossRef]
2. Watson, C.; Schalatek, L. *The Global Climate Finance Architecture*; Heinrich Böll Foundation: Washington, DC, USA, 2020.
3. Wong, G.Y.; Luttrell, C.; Loft, L.; Yang, A.; Pham, T.T.; Naito, D.; Assembe-Mvondo, S.; Brockhaus, M. Narratives in REDD+ Benefit Sharing: Examining Evidence within and beyond the Forest Sector. *Clim. Policy* **2019**, *19*, 1038–1051. [CrossRef]
4. Skutsch, M.; Turnhout, E. REDD+: If Communities are the Solution, What is the Problem? *World Dev.* **2020**, *130*, 1–9. [CrossRef]
5. Asiyanbi, A.P.; Ogar, E.; Akintoye, O.A. Complexities and Surprises in Local Resistance to Neoliberal Conservation: Multiple Environmentalities, Technologies of the Self and the Poststructural Geography of Local Engagement with REDD+. *Polit. Geogr.* **2019**, *69*, 128–138. [CrossRef]
6. Dunlop, T.; Corbera, E. Incentivizing REDD+: How Developing Countries are Laying the Groundwork for Benefit-sharing. *Environ. Sci. Policy* **2016**, *63*, 44–54. [CrossRef]
7. Awono, A.; Somorin, O.A.; Eba'a Atyi, R.; Levang, P. Tenure and Participation in Local REDD+ Projects: Insights from Southern Cameroon. *Environ. Sci. Policy* **2014**, *35*, 76–86. [CrossRef]
8. Pandit, R.; Neupane, P.R.; Wagle, B.H. Economics of Carbon Sequestration in Community Forests: Evidence from REDD+ Piloting in Nepal. *J. For. Econ.* **2017**, *26*, 9–29. [CrossRef]
9. Maraseni, T.N.; Neupane, P.R.; Lopez-Casero, F.; Cadman, T. An Assessment of the Impacts of the REDD+ Pilot Project on Community Forests User Groups (CFUGs) and Their Community Forests in Nepal. *J. Environ. Manag.* **2014**, *136*, 37–46. [CrossRef]
10. Andersson, K.P.; Smith, S.M.; Alston, L.J.; Duchelle, A.E.; Mwangi, E.; Larson, A.M.; de Sassi, C.; Sills, E.O.; Sunderlin, W.D.; Wong, G.Y. Wealth and the Distribution of Benefits from Tropical Forests: Implications for REDD+. *Land Use Policy* **2018**, *72*, 510–522. [CrossRef]

11. Chomba, S.; Kariuki, J.; Lund, J.F.; Sinclair, F. Roots of Inequity: How the Implementation of REDD+ Reinforces Past Injustices. *Land Use Policy* **2016**, *50*, 202–213. [CrossRef]
12. Samndong, R.; Vatn, A. Competing Tenures: Implications for REDD+ in the Democratic Republic of Congo. *Forests* **2018**, *9*, 662. [CrossRef]
13. Kansanga, M.M.; Luginaah, I. Agrarian Livelihoods under Siege: Carbon Forestry, Tenure Constraints and the Rise of Capitalist Forest Enclosures in Ghana. *World Dev.* **2019**, *113*, 131–142. [CrossRef]
14. Cavanagh, C.J.; Vedeld, P.O.; Trædal, L.T. Securitizing REDD+? Problematizing the Emerging Illegal Timber Trade and Forest Carbon Interface in East Africa. *Geoforum* **2015**, *60*, 72–82. [CrossRef]
15. Bayrak, M.M.; Marafa, L.M. Ten Years of REDD+: A Critical Review of the Impact of REDD+ on Forest-dependent Communities. *Sustainability* **2016**, *8*, 620. [CrossRef]
16. Froese, R.; Schilling, J. The Nexus of Climate Change, Land Use, and Conflicts. *Curr. Clim. Chang. Rep.* **2019**, *5*, 24–35. [CrossRef]
17. Scheffran, J.; Link, P.M.; Schilling, J. Theories and Models of Climate-Security Interaction: Framework and Application to a Climate Hot Spot in North Africa. In *Climate Change, Human Security and Violent Conflict: Challenges for Societal Stability*; Scheffran, J., Brzoska, M., Brauch, H.G., Link, P.M., Schilling, J., Eds.; Hexagon Series on Human and Environment Security and Peace; Springer: Berlin, Germany, 2012; Volume 8, pp. 91–131.
18. Seter, H.; Theisen, O.M.; Schilling, J. All about Water and Land? Resource-Related Conflicts in East and West Africa Revisited. *GeoJournal* **2018**, *83*, 169–187. [CrossRef]
19. Schilling, J.; Schilling-Vacaflor, A.; Flemmer, R.; Froese, R. A Political Ecology Perspective on Resource Extraction and Human Security in Kenya, Bolivia and Peru. *Extr. Ind. Soc.* **2020**, 1–12. [CrossRef]
20. Turner, M.D. Political Ecology and the Moral Dimensions of "Resource Conflicts": The Case of Farmer-Herder Conflicts in the Sahel. *Political Geogr.* **2004**, *23*, 863–889. [CrossRef]
21. Watts, M. Political Ecology. In *A Companion to Economic Geography*; Sheppard, E., Barnes, T., Eds.; Blackwell: London, UK, 2000; pp. 257–274.
22. Schilling, J.; Saulich, C.; Engwicht, N. Introduction: A Local to Global Perspective on Resource Governance and Conflict. *Confl. Secur. Dev.* **2018**, *18*, 433–461. [CrossRef]
23. Ide, T.; Lopez, M.R.; Fröhlich, C.; Scheffran, J. Pathways to Water Conflict during Drought in the MENA Region. *J. Peace Res.* **2021**, *58*, 568–582. [CrossRef]
24. Eisenhardt, K.M.; Graebner, M.E. Theory Building from Cases: Opportunities and Challenges. *Acad. Manag. J.* **2007**, *50*, 25–32. [CrossRef]
25. Blum, M. Whose Climate? Whose Forest? Power Struggles in a Contested Carbon Forestry Project in Uganda. *For. Policy Econ.* **2020**, *115*, 102137. [CrossRef]
26. Cavanagh, C.; Benjaminsen, T.A. Virtual Nature, Violent Accumulation: The 'Spectacular Failure' of Carbon Offsetting at a Ugandan National Park. *Geoforum* **2014**, *56*, 55–65. [CrossRef]
27. Massarella, K.; Sallu, S.M.; Ensor, J.E.; Marchant, R. REDD+, Hype, Hope and Disappointment: The Dynamics of Expectations in Conservation and Development Pilot Projects. *World Dev.* **2018**, *109*, 375–385. [CrossRef]
28. Scheba, A. Market-Based Conservation for Better Livelihoods? The Promises and Fallacies of REDD+ in Tanzania. *Land* **2018**, *7*, 119. [CrossRef]
29. Kemerink-Seyoum, J.S.; Tadesse, T.M.; Mersha, W.K.; Duker, A.E.C.; De Fraiture, C. Sharing Benefits or Fueling Conflicts? The Elusive Quest for Organizational Blue-prints in Climate Financed Forestry Projects in Ethiopia. *Glob. Environ. Chang.* **2018**, *53*, 265–272. [CrossRef]
30. Galudra, G.; van Noordwijk, M.; Agung, P.; Suyanto, S.; Pradhan, U. Migrants, Land Markets and Carbon Emissions in Jambi, Indonesia: Land Tenure Change and the Prospect of Emission Reduction. *Mitig. Adapt. Strateg. Glob. Chang.* **2013**, *19*, 715–731. [CrossRef]
31. Holmes, I.; Potvin, C.; Coomes, O. Early REDD+ Implementation: The Journey of an Indigenous Community in Eastern Panama. *Forests* **2017**, *8*, 67. [CrossRef]
32. Hoang, C.; Satyal, P.; Corbera, E. 'This is my Garden': Justice Claims and Struggles over Forests in Vietnam's REDD+. *Clim. Policy* **2018**, *19*, S23–S35. [CrossRef]
33. Sanders, A.J.P.; Hyldmo, H.D.; Prasti, H.R.D.; Ford, R.M.; Larson, A.M.; Keenan, R.J. Guinea Pig or Pioneer: Translating Global Environmental Objectives Through to Local Actions in Central Kalimantan, Indonesia's REDD+ Pilot Province. *Glob. Environ. Change-Hum. Policy Dimens.* **2017**, *42*, 68–81. [CrossRef]
34. Milne, S.; Mahanty, S. Value and Bureaucratic Violence in the Green Economy. *Geoforum* **2019**, *98*, 133–143. [CrossRef]
35. Pasgaard, M. Lost in translation? How Project Actors Shape REDD+ Policy and Outcomes in Cambodia. *Asia Pac. Viewp.* **2015**, *56*, 111–127. [CrossRef]
36. Lounela, A. Contested Values and Climate Change Mitigation in Central Kalimantan, Indonesia. *Soc. Anthropol.* **2020**, *28*, 862–880. [CrossRef]
37. Scheba, A.; Rakotonarivo, O.S. Territorialising REDD plus: Conflicts over Market-based Forest Conservation in Lindi, Tanzania. *Land Use Policy* **2016**, *57*, 625–637. [CrossRef]
38. Luttrell, C.; Loft, L.; Fernanda Gebara, M.; Kweka, D.; Brockhaus, M.; Angelsen, A.; Sunderlin, W.D. Who Should Benefit from REDD+? Rationales and Realities. *Ecol. Soc.* **2013**, *18*, 1–18. [CrossRef]

39. McDermott, M.; Mahanty, S.; Schreckenberg, K. Examining Equity: A Multidimensional Framework for Assessing Equity in Payments for Ecosystem Services. *Environ. Sci. Policy* **2013**, *33*, 416–427. [CrossRef]
40. Holmes, I.; Potvin, C. Avoiding Re-inventing the Wheel in a People-centered Approach to REDD+. *Conserv. Biol.* **2014**, *28*, 1380–1393. [CrossRef]
41. Agrawal, A.; Nepstad, D.; Chhatre, A. Reducing Emissions from Deforestation and Forest Degradation. *Annu. Rev. Environ. Resour.* **2011**, *36*, 373–396. [CrossRef]
42. Ribot, J.; Larson, A.M. Reducing REDD Risks: Affirmative Policy on an Uneven Playing Field. *Int. J. Commons* **2012**, *6*, 233–254. [CrossRef]
43. Oyono, P.R.; Ribot, J.C.; Larson, A. *Green and Black Gold in Rural Cameroon: Natural Resources for Local Governance, Justice and Sustainability*; Food and Agriculture Organization of the United Nations: Rome, Italy, 2006; Volume 22.
44. Sikor, T.; Nguyen, T.Q. Why May Forest Devolution Not Benefit the Rural Poor? Forest Entitlements in Vietnam's Central Highlands. *World Dev.* **2007**, *35*, 2010–2025. [CrossRef]
45. Larson, A.M.; Soto, F. Decentralization of Natural Resource Governance Regimes. *Annu. Rev. Environ. Resour.* **2008**, *33*, 213–239. [CrossRef]
46. Schreckenberg, K.; Luttrell, C. Participatory Forest Management: A Route to Poverty Reduction? *Int. For. Rev.* **2009**, *11*, 221–238. [CrossRef]
47. Bluffstone, R.; Robinson, E.; Guthiga, P. REDD+ and Community-Controlled Forests in Low-Income Countries: Any Hope for a linkage? *Ecol. Econ.* **2013**, *87*, 43–52. [CrossRef]
48. Awung, N.S.; Marchant, R. Quantifying Local Community Voices in the Decision-making Process: Insights from the Mount Cameroon National Park REDD+ Project. *Environ. Sociol.* **2018**, *4*, 235–252. [CrossRef]
49. Sills, E.; Atmadja, S.; Sassi, C.d.; Duchelle, A.E.; Kweka, D.; Resosudarmo, I.A.P.; Sunderlin, W. *REDD+ on the Ground: A Case Book of Subnational Initiatives Across the Globe*; CIFOR: Bogor, Indonesia, 2014.
50. Sunderlin, W.D.; Larson, A.M.; Duchelle, A.E.; Resosudarmo, I.A.P.; Huynh, T.B.; Awono, A.; Dokken, T. How are REDD+ Proponents Addressing Tenure Problems? Evidence from Brazil, Cameroon, Tanzania, Indonesia, and Vietnam. *World Dev.* **2014**, *55*, 37–52. [CrossRef]
51. Dokken, T.; Caplow, S.; Angelsen, A.; Sunderlin, W. Tenure Issues in REDD+ Pilot Project Sites in Tanzania. *Forests* **2014**, *5*, 234–255. [CrossRef]
52. Peters, E.P. Inequality and Social Conflict Over Land in Africa. *J. Agrar. Chang.* **2004**, *4*, 269–314. [CrossRef]
53. Lund, C. *Local Politics and the Dynamics of Property in Africa*; Cambridge University Press: Cambridge, UK, 2008.
54. Peluso, N.L.; Vandergeest, P. Political Ecologies of War and Forests: Counterinsurgencies and the Making of National Natures. *Ann. Assoc. Am. Geogr.* **2011**, *101*, 587–608. [CrossRef]
55. Cotula, L.; Mayers, J. *Tenure in REDD: Start-Point or Afterthought*; IIED: London, UK, 2009.
56. Larson, A.M. Forest Tenure Reform in the Age of Climate Change: Lessons for REDD+. *Glob. Environ. Chang.* **2011**, *21*, 540–549. [CrossRef]
57. Naughton-Treves, L.; Wendland, K. Land Tenure and Tropical Forest Carbon Management. *World Dev.* **2014**, *55*, 1–6. [CrossRef]
58. Corbera, E.; Estrada, M.; May, P.; Navarro, G.; Pacheco, P. Rights to Land, Forests and Carbon in REDD+: Insights from Mexico, Brazil and Costa Rica. *Forests* **2011**, *2*, 301–342. [CrossRef]
59. Duchelle, A.E.; Cromberg, M.; Gebara, M.F.; Guerra, R.; Melo, T.; Larson, A.; Cronkleton, P.; Börner, J.; Sills, E.; Wunder, S.; et al. Linking Forest Tenure Reform, Environmental Compliance, and Incentives: Lessons from REDD+ Initiatives in the Brazilian Amazon. *World Dev.* **2014**, *55*, 53–67. [CrossRef]
60. Murdiyarso, D.; Brockhaus, M.; Sunderlin, W.D.; Verchot, L. Some Lessons Learned from the First Generation of REDD+ Activities. *Curr. Opin. Environ. Sustain.* **2012**, *4*, 678–685. [CrossRef]
61. Phelps, J.; Webb, E.L.; Agrawal, A. Land Use-Does REDD+ Threaten to Recentralize Forest Governance? *Science* **2010**, *328*, 312–313. [CrossRef] [PubMed]
62. Sandbrook, C.; Nelson, F.; Adams, W.M.; Agrawal, A. Carbon, Forests and the REDD Paradox. *Oryx* **2010**, *44*, 330–334. [CrossRef]
63. Hayes, T.; Persha, L. Nesting Local Forestry Initiatives: Revisiting Community Forest Management in a REDD+ World. *For. Policy Econ.* **2010**, *12*, 545–553. [CrossRef]
64. Cronkleton, P.; Bray, D.B.; Medina, G. Community Forest Management and the Emergence of Multi-Scale Governance Institutions: Lessons for REDD+ Development from Mexico, Brazil and Bolivia. *Forests* **2011**, *2*, 451–473. [CrossRef]
65. Adhikari, S.; Kingi, T.; Ganesh, S. Incentives for Community Participation in the Governance and Management of Common Property Resources: The Case of Community Forest Management in Nepal. *For Policy Econ* **2014**, *44*, 1–9. [CrossRef]
66. UN. U.N. Report of the Conference of the Parties on its Sixteenth Session, Held in Cancun from 29 November to 10 December 2010. Available online: https://www.google.com/url?sa=t&rct=j&q=&esrc=s&source=web&cd=&ved=2ahUKEwiEstrMpcnwAhVF2qQKHfySB8YQFjAAegQIBhAD&url=https%3A%2F%2Funfccc.int%2Fresource%2Fdocs%2F2010%2Fcop16%2Feng%2F07a01.pdf&usg=AOvVaw36AirtEfa7AMHGXVZzKMNj (accessed on 14 May 2021).
67. Kemp, D.; Owen, J.R. Corporate Readiness and the Human Rights Risks of Applying FPIC in the Global Mining Industry. *Bus. Hum. Rights J.* **2017**, *2*, 163–169. [CrossRef]

Short Biography of Authors

Janpeter Schilling is the Scientific Director of the Peace Academy Rhine-land-Palatinate. He also holds a Klaus Töpfer Junior Professorship for Landuse Con-flicts at the University of Koblenz-Landau. He received his Phd in Geography in 2012 from the University of Hamburg. He has worked at universities and peacebuilding or-ganizations in Hamburg, Berlin, London, and New York. His research focuses on con-flict, resources, and environmental change.

Rowan Alumasa Alusiola is currently, an Alexander Von Humboldt Fellow for International Climate Protection at Koblenz University- Landau. She is finalizing her master's degree program in Environmental Studies and Community Development at Kenyatta University in Kenya. She has 10 years of practical work experience with both National and International NGOs in Program development and management. Her re-search focuses on forest conservation, conflict, and local community's livelihood.

Paul Klär is currently working on his master's degree in science communication at the Karlsruhe Institute of Technology. He has 3 years of experience as a research assistant in different fields at University of Koblenz-Landau and the Peace Academy Rhineland-Palatinate. In his bachelor's thesis he worked on conflicts about access to water in Africa.

forests

MDPI

Article

Stakeholders' Perception of the Impact of the Declaration of New Protected Areas on the Development of the Regions Concerned, Case Study: Czech Republic

Jiří Schneider, Aleš Ruda * and Martina Blahová

Department of Environmentalistics and Natural Resources, Faculty of Regional Development and International Studies, Mendel University in Brno, tř. Gen. Píky 7, 61300 Brno, Czech Republic; jiri.schneider@mendelu.cz (J.S.); zaneta.kalasova@mendelu.cz (M.B.)
* Correspondence: ales.ruda@mendelu.cz; Tel.: +420-545-136-252

Abstract: Floodplain forests at the confluence of the rivers Dyje and Morava (in the southeastern tip of the Czech Republic) are completely unique ecosystems in terms of area and ecology. For many years, there has been an effort by the state's nature protection officials to declare the area as a Protected Landscape Area. This effort is met by the resistance of foresters and other local stakeholders. The study focuses on the identification of stakeholders' comments and objections to the planned declaration of the Soutok PLA and the comparison between the objections raised and the attitudes of stakeholders from existing PLAs. Using the content analysis of 247 paper documents, the first part of the study determines the negative arguments that are subsequently verified in the second part on the basis of 17 semistructured standardized interviews and interview surveys of 200 respondents. The analysis of the interviews and surveys was based on the grounded theory method. The theoretical sampling and snowball techniques were used to recruit the respondents. The interviews and surveys showed that most concerns over restrictions established by the conservation status are unnecessary since experience showed that they are either not registered or not established by the PLA status, and their application is provided by other legislative standards.

Keywords: protected areas establishment; stakeholder participation; landscape protection; qualitative research; Soutok Protected Landscape Area (Czech Republic)

Citation: Schneider, J.; Ruda, A.; Blahová, M. Stakeholders' Perception of the Impact of the Declaration of New Protected Areas on the Development of the Regions Concerned, Case Study: Czech Republic. *Forests* **2021**, *12*, 580. https://doi.org/10.3390/f12050580

Academic Editors: Juan F. Fernández-Manjarrés and Roxane Sansilvestri

Received: 22 March 2021
Accepted: 28 April 2021
Published: 6 May 2021

1. Introduction

Human activities, including agriculture, forestry, and urbanization, have influenced ecosystems to such an extent that there is only a small number of natural places in the world left [1]. As a result, protected areas are considered crucial for nature conservation [2] and the maintenance of biodiversity. The protection of natural areas is presented as a useful tool for many reasons, the most important being provisioning, regulating, cultural, and supporting ecosystem services [3,4].

Protected Landscape Areas (PLAs) are one of the categories of protected areas (PAs) according to Act 114/1992 Coll. on nature and landscape protection. PLAs are defined as large areas with a harmoniously shaped landscape, characteristic relief, a significant proportion of natural forest ecosystems, and permanent grasslands. The economic use is carried out according to the zones of graded protection so that their natural state is maintained and improved, and the optimal ecological functions of these areas are preserved or recreated. The protection regime is therefore looser here than in the case of national parks.

Social support is one of the key factors that determine the successful establishment and management of protected areas [5–8]. The perception of a Protected Landscape Area (PLA) in designation could differ from the perception of an existing Protected Landscape Area (PLA), which has been operating for years, where perception is influenced by experiences with management. Only a few studies explored perceptions before the establishment of

PLAs [9,10] even though knowledge about them is actually most important for the success of PLA establishment [11].

Currently, nature and landscape protection are much-debated issues, especially in the context of regional development. Declaration of large-size protected areas often leads to controversial and emotional debates. The authors highlighted three reasons why designated areas may not reflect public perceptions. These include the following reasonings: (i) designated landscapes may be based on historical decisions made before participatory processes were included; (ii) expert knowledge is often the only form of knowledge considered or tends to be privileged over other forms of knowledge; and (iii) the focus has been on certain landscape evaluation criteria over others (e.g., recreational, therapeutic, and spiritual values, sense of place or scenic qualities) [12]. While conservation strategies typically account for changing forest ecology, wildlife populations, and biophysical conditions, far less consideration is given to the changing perceptions of PLA neighbors [13,14].

On the other hand, in many cases, conservation activities have resulted in positive impacts for local communities, including livelihood provision through tourism development, economic benefits, environmental management, participation in governance, and the protection of historic sites and cultural resources [15–17]. If the set of management actions lived up to people's expectations, protected areas would have a higher probability of being socially supported, which must be a central point in their adaptive management [8]. The attractiveness of the preserved state of nature and landscape, which is in the long term guaranteed by the status of the protected area, is undoubtedly one of the important prerequisites for the long-term economic prosperity of local municipalities and subregions.

Rather than simply labeling conservation as positive or negative, actions taken to protect or manage the environment can produce a suite of both positive and negative impacts that variably affect social, economic, cultural, health, and governance spheres of local communities [17]. Conservation strategies are currently evolving from wilderness protection and restoration to the appropriate use of natural resources and maintenance of landscapes influenced by human management [18]. According to Bennett [19], local people's perceptions about conservation in PAs can be categorized into four thematic areas: social impacts of conservation, ecological outcomes of conservation, legitimacy of conservation governance, and acceptability of conservation management. This kind of categorization could be helpful in identifying aspects of conservation policies and management actions that are acceptable or unacceptable to local people [20].

Perception and attitudes toward protected areas establish the degree of success when it comes to sustainable conservation planning [21] because conservation initiatives require the active participation of local communities in decision-making processes and solutions to integrate local development with environmental conservation [21]. Local community involvement in the management of protected areas usually leads to increased awareness of the benefits of biodiversity, more responsible use of resources, and welfare of local people [22].

In the past decades, there has been a shift in the perception of protected areas in post-communist countries. The conventional method (officially ordered nature conservation without the acceptance of stakeholders' opinion) is abandoned and the participatory method of nature and landscape conservation is becoming the center of attention [11]. In contemporary literature, local residents' perceptions and attitudes toward protected areas are identified as key factors for the successful management of these areas [23]. According to Nastran [11], the main factors influencing the stakeholders' attitude to future protected areas are the perception of benefits, effects and costs, the possibility to participate in important meetings about the protected area, and personal experience with previous negotiations with nature conservation authorities. According to Winter et al. [10], the perception of protected areas is further influenced by socioeconomic factors (age, sex, education, ethnic origin, place of residence, etc.). Trakolis [24] stated that it is necessary to consider the divergence of opinions and the perception of benefits of the protected area between stakeholders and administrators (founders). Allendorf et al. [25] also pointed out that some one-off or past actions are quickly forgotten by stakeholders (mainly infrastructure projects and increased employment in services). Such a situation when people forget that protected areas generate benefits given by ecosystem services significantly contributes to the rather negative perception of protected areas [26]. The existing problems with the protection of nature highlight the need for a transdisciplinary approach. Natural areas can no longer be protected by normative regulations in situ but social factors that strongly influence the success of declaring a protected area have to be included [11].

In addition to specially protected areas, much attention has recently been paid to the perception of stakeholders at Natura 2000. It is also well known in the Soutok area. Apostolopoulou et al. [27] carried out an analysis of the participatory process in the management of the Greek Natura 2000 sites, while Rojas-Briales [28] investigated the key socioeconomic issues in the implementation of the Natura 2000 network in Spain or Nastran, and Pirnat [29] showed the stakeholder involvement in the designation of the Natura 2000 sites in Slovenia [30].

The main objective of the study is to determine the stakeholders' negative attitude factors against the declaration of the Soutok PLA and to verify the adequacy of the arguments raised by the stakeholders (representatives of local authorities, the public, interested and professional organizations), which were used as a negative stance against the declaration of the Soutok PLA. During the research, we tested the following working hypothesis: Stakeholders prevent the emergence of PLAs from irrational fears and stereotypes, or ignorance (and generally insufficient environmental education and awareness in the Czech Republic) or from their own selfish interests. Most, (in fact) almost all, of their arguments are unfounded.

2. Materials and Methods

2.1. Study Area

The Czech Republic is a landlocked hilly plateau surrounded by relatively low mountains, including the Carpathian, Ore, Sudetic, and the Sumava mountains, and lies in the temperate climate zone with average temperatures varying among the various regions of the Czech Republic; it is also characterized by mild and humid summers with occasional hot spells, and cold, cloudy, and humid winters. The Soutok (confluence, in English) area (Figure 1) is a unique landscape in the downstream parts of the Morava and Dyje rivers [31] with a mosaic of forest, meadow, wetland, and water ecosystems, with a large complex of lowland floodplain forests and part of unique historical landscaping in the Lednice-Valtice Area (UNESCO World Heritage Site and also part of UNESCO Biosphere reserve Dolní Morava). A substantial part has the character of primeval stands that, together with other types of biotopes, form an indivisible complex (Figures 2–7). The area is also extremely important for the size and quality of meadow habitats, especially continental flooded meadows. The Ministry of the Environment of the Czech Republic demarcated the Soutok PLA into two discontinuous blocks with a total area of 139 km^2, separated by the town of Břeclav. The V-shaped area spreads along the Morava and Dyje rivers starting at their upstream confluence. The PLA extends to the cadastral territories of 19 municipalities in the districts of Břeclav and Hodonín. Both districts are border regions that are developing cooperation with the municipalities of Lower Austria and western Slovakia, especially in the construction of transport infrastructure and environmental protection. The area is defined and delineated so that it does not extend to the built-up parts of the municipalities.

The existing protected landscape areas of Moravský Kras/Moravian Karst and Pálava/Pavlov Highlands (Figure 1) were chosen to verify the stakeholders' arguments against the declaration of the Soutok Protected Landscape Area. Both territories are located in the South Moravian Region. Moravský Kras/Moravian Karst was declared a Protected Landscape Area in 1956 as the second PLA in the territory of the present-day Czech Republic. Pálava/Pavlov was declared a Protected Landscape Area in 1967.

Moravský Kras (Moravian Karst) is a 3 to 6 km wide and 25 km long strip of Devonian limestone [32]. The long-term goal of nature and landscape protection in the PLA is to preserve the unique set of surface and underground karst phenomena as the basis of the typical landscape character [33]. The Moravian Karst was declared a PLA covering an area of 92 km^2 as early as 1956, which made it the second oldest PLA in the present-day Czech Republic. The value of the area was reflected in the high representation of the most strictly protected sites. Zone 1 occupies 17.7% of the PLA area.

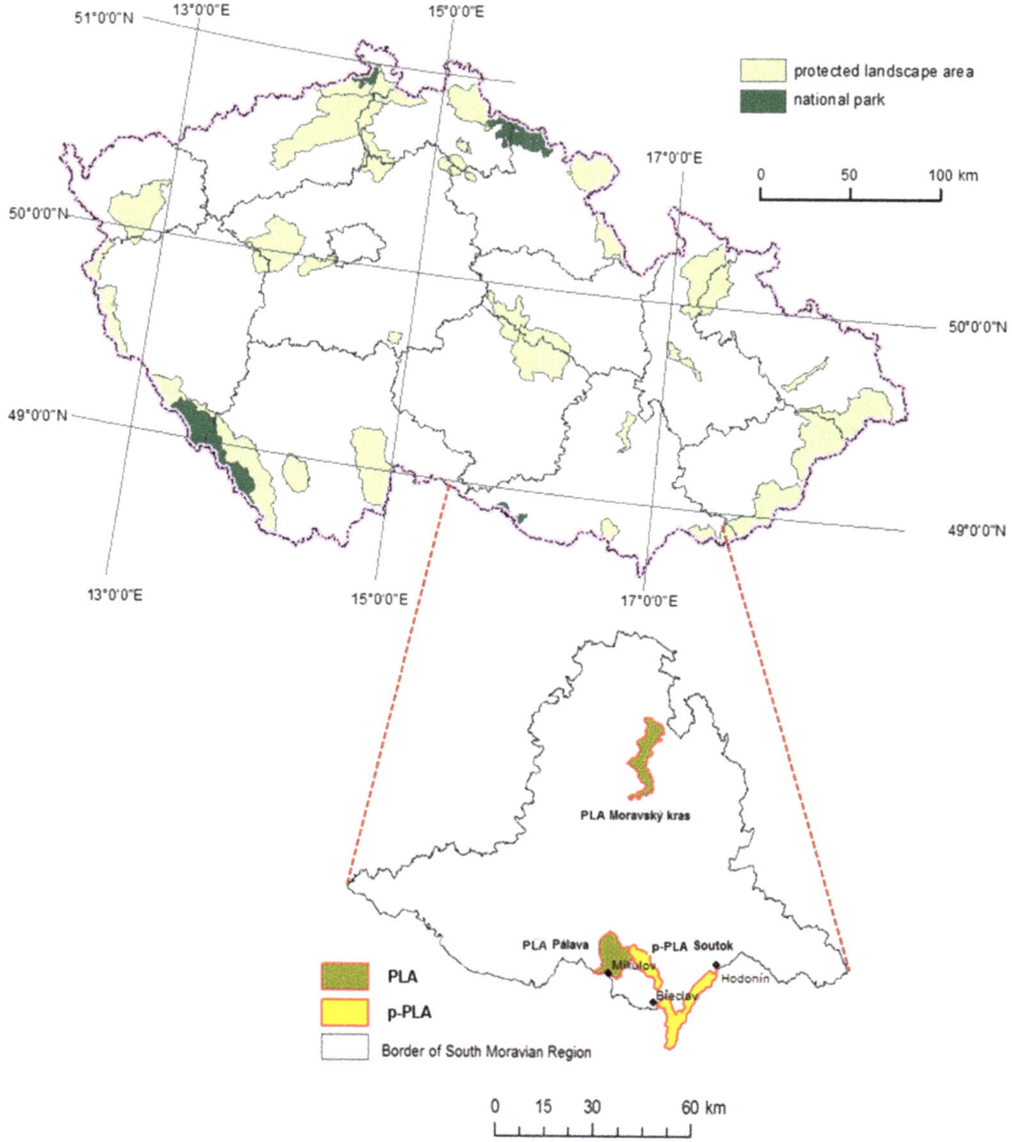

Figure 1. Study area (p-PLA planned protected landscape area).

Figure 2. The floodplain forests in the Soutok area are interwoven with a network of river branches, billabongs, canals, etc.

Figure 3. Which are the habitats of many specially protected plants and animals (e.g., *Rana* sp.).

Figure 4. The mosaic of floodplain habitats is also cocreated by Willow–poplar forests of lowland rivers, tall-sedge beds, Eutrophic vegetation of muddy substrates, etc.

Figure 5. The most valuable parts are of primeval character with a lot of woody debris.

Figure 6. In addition to forests, the landscape of Soutok is formed by Continental inundated meadows with solitaires of living or dead trees, the niches of which are inhabited by specially protected animal species (e.g., white stork).

Figure 7. The result of the clear-felling system of forest management is even-aged and monocultural stands of deciduous, mostly autochthonous, tree species (oak, ash).

The area of the Protected Landscape Area Moravský Kras is located in cadastral areas of 24 municipalities. However, only the cadastral areas of the municipalities of Ostrov u Macochy and Rudice overlap completely with this protected landscape area. According to the data of the Czech Statistical Office, as of 1 January 2015, the population was 24,850 inhabitants. For the share of economic subjects in chosen administrative areas of municipalities with extended powers on the territory of protected landscape areas, see Table 1.

Table 1. The share of economic subjects in chosen administrative areas of municipalities with extended powers on the territory of protected landscape areas of Moravský Kras and Pálava, 2015 (Source: CZSO).

Share of Economic Subjects Based on Chosen Sectors of Economic Activities (%)	Protected Landscape Area Pálava		Protected Landscape Area Moravský Kras		
	Municipality with Extended Powers Břeclav	Municipality with Extended Powers Mikulov	Municipality with Extended Powers Blansko	Municipality with Extended Powers Brno	Municipality with Extended Powers Šlapanice
Agriculture, Forestry, and Fishing	7.6	10.6	4.1	1.8	4.1
Industry total	12.2	13.8	16.1	10.4	15.8
Construction	11.2	15.3	11.2	8.6	12.0
Retail, Accommodation, Meals and Restaurant Services	27.4	26.5	24.5	27.3	24.8

The Pálava PLA (Pavlov Highlands) covers an area of 83 km^2, making it one of the smallest PLAs in the Czech Republic [34]. Overall, 30% of its area is covered by forests, mainly in Děvín and the eastern part of the territory, and 55% is agricultural land [32] with traditional vineyards. It is a particularly valuable biogeographical site with thermophilic ecosystems of almost Mediterranean character. Floodplain forests, wetland meadows in the Dyje floodplain, and one of the last habitats of salinephilic vegetation—Slanisko Nesyt—increase the diversity of the area. The character of the landscape is completed by agriculturally used plots with the predominance of vineyards, and individual settlements with the privileged position of the historic town of Mikulov [35].

According to the data of the Czech Statistical Office, as of 1 January 2015, the population was 13,344 inhabitants in the 11 cadastral areas. From this number, more than half of the inhabitants live in the municipality Mikulov (7443). The number of inhabitants in the remaining municipalities does not exceed 1000, which makes this region a typical rural area.

2.2. Data Collection and Analysis

The data were collected in two stages (Figure 8). The first stage included the collection of material for the identification of key objections and comments that the stakeholders (representatives of municipalities, economic operators, and individual citizens) used to argue against the declaration of the Soutok PLA for fear of complications that the new nature protection scheme would bring. A total of 93 min of meetings of municipal councils, 86 newspaper articles and special reports from environmental bodies, 19 decisive stances of municipalities, and 49 stances of the economic operators concerned were collected from publicly available sources and databases of the municipalities concerned during the 2008–2013 negotiations. The documents were subsequently submitted to content analysis, with the help of which we determined and categorized the key arguments of the stakeholders. Content analysis represents one of the techniques of document analysis that aims to organize and reveal latent information in written materials. It is based on an objective, systematic, and quantitative description of the content of the communication [36]. For qualitative content analysis, we used Mayring's approach [37,38], working with coding criteria that are developed based on research questions and theoretical background, while categories are derived inductively from the collected material. Categories are also verified and refined in order to fit previously established criteria. When examining the content, content units and categories were identified first and then used to evaluate the negative arguments in the final analysis. During the coding and classification process, each part was supervised and discussed by the authors in order to achieve relevant results. Depending on the chosen methodology, inductive or deductive coding or both might be used. Inductive coding starts with a detailed analysis of sources (articles, concepts, themes, etc.) and is useful for strategies using grounded theory. On the other hand, deductive coding works with a set of keywords, specific themes, and ideas, followed by indicating whether these are mentioned in the sources [39]. It enabled us to determine both the quantity of text expression (amount and frequency) and also the quality of new themes. Using deductive category application, the sources are analyzed according to a coding framework, and we can use different techniques: formal and content structuring, typifying, and scaling structuring [40]. For our study, we used a combination of inductive and deductive coding approaches. Generally, we can sum up deductive–inductive coding process into two steps. In the first step, we developed deductive categories, as mentioned above, and in the second step, we developed inductive categories for the analysis and the interpretation of new aspects and themes.

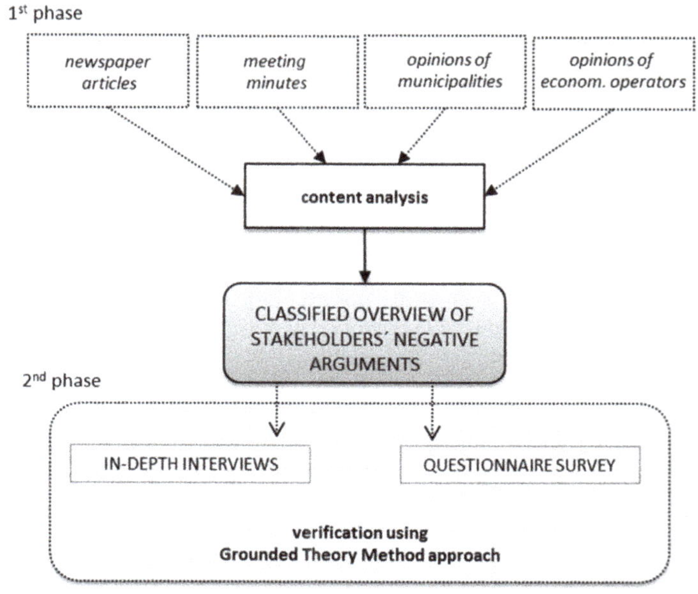

Figure 8. Methodological framework.

In this study, we used NVivo 11 software to conduct the content analysis. NVivo offers researchers the possibility to import various types of raw data (text files, audio and video files, and images), and although it can run automatically, we used a software-assisted approach and checked the data processing by reviewing the word list, removing irrelevant words and rerunning the query until the created proposed model (with parent nodes and child nodes) fit the best to present the findings [41].

The second stage represented a confrontation of the negative arguments with the attitudes of the representatives of the municipalities located in the existing PLAs in the same region that have a similar or close subject of nature and landscape protection as the proposed Soutok PLA. These were the Moravian Karst PLA and the Pálava PLA. For these needs, materials for in-depth, semistructured, and open-ended interviews and questionnaires were assembled. Respondents for face-to-face interviews were chosen on the basis of the purpose-made selection method using the snowball technique, in which other respondents were recruited on recommendation from previous interviews. In each of the selected PLAs, we chose both mayors of municipalities situated on the outskirts, which are almost unaffected by tourism, and municipalities located near the most visited tourist destinations. Within the Moravian Karst PLA, these included the municipalities of Blansko, Kanice, Ostrov u Macochy, and Vilémovice, and in the case of the Pálava PLA, the municipalities of Dolní Věstonice, Pavlov, Přítluky, and Sedlec u Mikulova. Furthermore, interviews were conducted with the representatives of major forest owners—the School

Forest Enterprise Masaryk Forest Křtiny (Křtiny, Czech Republic) and the Židlochovice Forest Enterprise (LZ), the managers of PLA administration involved and other economic operators (AGRO Měřín Agricultural Holding). A total of 17 structured interviews with a standardized list of questions lasting 45–60 min were conducted in 2017. The arguments concerning the restrictions on the residents and visitors were evaluated on the basis of a questionnaire survey carried out in the summer months of 2018, i.e., during the main tourist season. We used the theoretical sampling technique [42] to select the respondents who answered partial questions within the scope of the Likert scale while openly commenting on partial arguments. It is based on a process of data collection for generating theory whereby the analyst jointly collects codes and analyses data and decides what data to collect next and where to find them. For the questionnaire survey, we used structured questionnaires that we administered personally and we collected 200 questionnaires with a response rate of 80%.

The methodological framework covered the first phase—investigation in the proposed Soutok PLA and the second phase—the verification within present PLA/LPAs Moravský Kras and Pálava (Figure 8).

The grounded theory method presented by Glaser and Straus [42] was used to evaluate the interview and questionnaire data. It is probably the most widely used approach in qualitative research [43]. By formulating a research question, it seeks to create theories firmly grounded in data, thus distinguishing itself from theories that are created in an intuitive or speculative manner. It is a method of data analysis and a comprehensive approach that governs the research process from finding a research question to publishing the resulting theory and its possible further development. We used this approach to discover the motives and hidden relations between the perception of negations caused by the declaration of the Soutok PLA. When applying this approach, we proceeded in the following steps: (1) creating concepts, (2) seeking theoretical relations between concepts, and (3) selecting a central concept and formulating the theory. When creating the concepts, we used open coding, then axial coding, and selective coding.

3. Results

Of the total number of 200 respondents, 108 women (54%) and 92 men (46%) participated in the questionnaire survey. The majority of respondents belonged to the age group of 18–39 (40.5%), 67 respondents (33.5%) to the age group of 40–59, and 26% of participants to the age group of over 60. In terms of the highest educational attainment, the respondents were divided into almost equal groups. We addressed 48 persons (24%) with lower-level secondary education and 71 persons (35.5%) with an advanced level of secondary education, attested by a certificate of apprenticeship or a diploma. A total of 27% of those surveyed had completed tertiary vocational education and 13.5% university education. Moreover, 102 persons living in the municipalities extending to the Moravian Karst PLA or the Pálava PLA were set aside by a filter question. Of these, 54 respondents live in the Moravian Karst PLA municipalities and 48 in the Pálava PLA municipalities.

3.1. Arguments against the Declaration of the Soutok PLA—Overview

Although negotiations on the declaration of the Soutok PLA have been held several times, it has not been enforced mainly due to the constant resistance of local municipalities and residents, whose negative attitude persists. Using contact analysis of available documents, we created a categorization of the negative arguments (Table 2) that the stakeholders repeatedly used to express their disapproval of the declaration of PLA. Nevertheless, even the opponents of the proposal are aware of the extraordinary value of the area. However, they are still convinced that the conservation status is sufficient and that the PLA would increase bureaucracy and restrictions. Opponents of the project include all mayors of the municipalities concerned, Forests of the Czech Republic, s.p. (state enterprise) on behalf of the Židlochovice Forest Enterprise, which is the administrator of most forests in the area, and the representatives of the Lower Morava Biosphere Reserve.

3.2. Arguments against the Declaration of the Soutok PLA—Increase in Stakeholders' Restrictions

Mainly, the residents of the affected municipalities were concerned about the restrictions on normal movement in the area of the proposed Soutok PLA. These doubts are based on past experience when the territory belonged to the border zone where entry was banned for 40 years. The residents were therefore afraid of further bans and restrictions that, in their words, the PLA status would bring.

Table 2. Objections and comments of stakeholders from the area concerned on the declaration of the Soutok PLA, own processing.

Category	Unit	Description	Status/Relevance
limitations	entry into and movement in PLA	residents' concerns over restrictions on entry and free movement based on Section 26 of Act No. 114/1992 Sb., on nature and landscape protection [44]	Not confirmed
	agricultural and forestery production	restrictions on agricultural and forestry production resulting from Act No. 114/1992 Sb., on nature and landscape protection [44]	Confirmed
	ship transportation	restrictions/termination of boat cruises on the Chateau Pond and on the Morava river from Břeclav to Janův hrad	Irrelevant
	compromised identity	alienation of the residents in relation to the protected area as a result of various prohibitions and restrictions, disregard of regional and local needs and customs	Not confirmed
	constructions	intervention of officials in built-up parts of the area in connection with the appearance of buildings and their color design	Confirmed
	bike trails construction	ban on the construction of new asphalt bike trails after the declaration of PLA; preservation of the existing trails is not guaranteed	Not confirmed
regional development	construction closure	restricted building work that would hamper the development of municipalities thus leading to depopulation and, most of all, the outflow of business operators	Confirmed
	slowing/stopping regional development	fear that the municipalities will be gripped by the protected areas without having a possibility to change the master plan toward the expansion of built-up areas	Not confirmed
	decline in cross-border cooperation	negative impact on cross-border cooperation, e.g., impossibility to build a footbridge across the Dyje river to the Austrian municipality of Rabensburg	Not documented. Estimate: Not confirmed
	decline in tourism	restricted development of tourism related to the restrictions on the movement of people in PLA	Not confirmed
	loss of job opportunities	loss of job opportunities, especially in agriculture and forestry that employ high numbers of people	Not confirmed
economy	decline in property prices	concerns over the decline of land and real estate prices due to regional development limits caused by the declaration of PLA	Not confirmed
	decrease in the number of small and medium-sized businesses	constraints to business development caused by limited construction of technical infrastructure and civic resulting from already low purchasing power of residents in rural areas	Not confirmed
	reduction in municipality income	under Act No. 114/1992 Sb. [44] mining of minerals in PLA zone 1 is prohibited, which will lead to a reduction in the present extraction of oil, natural gas, and gravel sand by private companies and thus to a reduction in the income of the municipalities	Not confirmed
	increase in investments connected with the declaration of PLA	the state does not provide any compensation for the increase in investment caused by the protection	Not confirmed

Table 2. *Cont.*

Category	Unit	Description	Status/Relevance
environmental protection	insufficient environmental management plan	the management plan is not drawn up pursuant to Section 1 of Decree No. 80/2008 Sb., on management plans, designation and registration of areas protected under the Nature and Landscape Protection Act	Not confirmed
	complications in adopting FMP	complications in drawing up the forest management plan (FMP) mean restrictions on harvesting forest in the area of interest and designing FMP that will primarily satisfy the Agency for Nature Conservation and Landscape Protection	Confirmed
	sufficient conservation status	the proposed area currently belongs to the Natura 2000 network, there is a biosphere reserve, a UNESCO site, and several small special protection areas that provide local protection, so general protection is unnecessary	Not confirmed
	our work, our care	the current state of the area is an achievement of forest managers from the Židlochovice Forest Enterprise, who have been farming here for several generations	Not confirmed
public administration and self-government	complications in municipal self-government	the nature conservation authority with competence for the territory directly in the PLA will be the PLA administration and not the municipal authority with enlarged jurisdiction or the regional authority, which will complicate the submission of permit applications	Not confirmed
	lack of information	the Ministry of the Environment provided incomplete and biased information about the declaration process of the Soutok PLA	Not confirmed

Timber production and terrain conditions of forests (slopes, field accessibility) play an important role in understanding the conflicts between nature conservation and forest management. Floodplain forests of the proposed Soutok PLA represent the most productive forest ecosystems in the Czech Republic.

In the short term, the introduction of close to nature management may partially increase the cost of forest management, but the use of natural processes in forests brings a significant financial benefit in the long term [45]. In forest management, the use of forests should not bring just a short-term economic benefit. The widespread use of natural processes in forestry practice has a profound and long-term biological impact on the prosperity of forest ecosystems and thus the economy of their management.

Research has revealed that the status of PLA is not related to the reduced attractiveness of the area. Overall, 127 respondents (63.5%) find areas outside large PLAs as attractive as national parks and PLAs themselves, and 70 respondents (35%) find large protected areas more attractive. The majority of respondents (73%) would be equally interested in the site even before the declaration. Most of them argued that "the declaration will not change the area in any way" and that "the site is either interesting or not, and this is not related to the PLA status." Furthermore, 26% of the respondents will find the site more attractive because "there had to be a reason for the declaration" or "we will get better information about the area" and "the site will be better mapped." Only two of the respondents would find the site less attractive due to the fact that there will be a "less interesting nature" and also because these areas are "supervised," which seemed to be a limiting factor to them.

Representatives of municipal authorities unanimously agreed that certain rules for the shape of houses, colors of roofs, etc. are defined. However, not all respondents consider this to be a limitation. The most frequent limitation they mentioned was the duty to build gable roofs with red to brown burnt tiles and the impossibility to have dormers and balconies. Overall, 73.5% of participants did not feel restricted in the construction of the house.

3.3. Arguments against the Declaration of the Soutok PLA—Negative Impact on the Development of the Region

Successful and sustainable development of the region is guaranteed by a well-designed master plan and balanced use of the potential of the area. In addition, the layout and land use are bound by generally applicable territorial limits. The responses of municipality mayors to the question of restrictions on construction in PLA are different and each of them sees restrictions from a different perspective. An important factor is also the location of the built-up part of the municipality in relation to the PLA. The Soutok PLA has been designed so that it does not extend to the built-up area. The proposal even featured a possible bypass of the city of Břeclav, when the area needed for the construction of new buildings was excluded from the PLA proposal in order to avoid potential conflict situations. The mayors of municipalities in current protected landscape areas do not think that PLAs limit the development of municipalities. In their words, entrepreneurs do not register offices in these municipalities as PLAs do not allow the construction of logistics centers or large development projects. However, this does not mean that the declaration of PLA is responsible for the declining number of small and medium-sized businesses. At the European level, there is a continuous need to strengthen relations between the environment, regional development, and regional planning. It is crucial to take into consideration the territorial impact on environmental planning, especially in the areas of planning and management of protected areas, water resources, soil erosion, or localization of dangerous and highly polluting substances, which are closely related to the impact on the landscape. The key concern should be the mobilization of the inner potential of municipalities in the form of brownfield revitalization rather than an extensive spatial extension. Similarly, the presence of PLA was not found to be a restriction on cross-border cooperation. Instead, the mayors emphasized the language barrier to be the main constraint.

An important question in the field of tourism is to what extent visitors affect the phenomena for which protection was declared. Any applicable restrictions are established in order to protect the phenomena that are destroyed by an onslaught of tourists. These are prohibitions and orders to guide tourists toward sustainable tourism so that there is still something to protect in the next decades. However, the interviews with the mayors showed that the municipalities are trying to develop tourism and attract tourists to other less well-known areas of the PLAs. Based on the research, the assumption of job losses related to the declaration of the PLA was not confirmed.

3.4. Arguments against the Declaration of the Soutok PLA—Economic Risks

Given the fact that PLAs were declared several decades ago, it is impossible to find out whether and how the declaration of PLA was reflected in the price of land. Comparing prices inside and outside PLAs seems to be a possible starting point, but it is very problematic since many more factors affect the price than just the PLA. Thus, mere price comparison cannot be considered a relevant methodology. For the above-given reasons, the exact land price movements cannot be determined. However, it is possible to point out low market intensity within the PLA, with few sales and offers made in comparison with the neighborhoods. Thus, only a fraction of what is sold outside is sold inside PLAs. Personal interviews showed that the respondents could not assess whether and how the land price had changed, precisely because the PLAs were declared a long time ago. Only one mayor pointed out that, in his opinion, the land inside the PLA was virtually unsaleable because of land use restrictions. An important argument for the municipalities in the proposed Soutok PLA was the possibility of extracting oil, natural gas, and gravel sand from which businesses contribute to the municipal budget. Partial objectives of the Moravian Karst PLA administration include a ban on further mining activities—not even in close proximity (currently only one surface limestone quarry is active), prevention of the establishment of new mineral deposits, and prevention of the declaration of new mining areas and support for restoration projects with maximum use of natural processes. This mention thus confirms the concern over the impossibility of mining in the newly declared PLA. On the other hand, the extraction of oil and natural gas does not take place on the surface in large mining areas but rather on spots. Real estate tax may pose another problem for the municipal budget. The land included in PLA zone 1 [46] is also exempt from real estate tax. Thus, municipalities lose part of the cadastral territory to this tax, which is otherwise fully paid to the municipality in whose territory the property is located [47]. It was often mentioned that nature conservation means increased investments, which must be paid from the municipal budget, but the state does not offer any compensation for them. However, municipality representatives mentioned that they can apply for subsidies, which in turn will relieve the municipal budget. It is precisely because of the location in the protected landscape area that the municipality is either awarded extra points for the location or receives a subsidy that can cover up to 100% of eligible costs. Interviews with representatives of municipalities clearly showed that nature conservation in the form of PLA does not mean an increase in investments for municipalities.

3.5. Environmental Protection

All existing territories in the Czech Republic have been cocreated or influenced by human beings and for their long-term functioning and biodiversity conservation other human interventions are usually necessary [48]. Although forest managers cannot be denied their share of the current state of forests in the area of the proposed Soutok PLA, the representative of the Židlochovice Forest Enterprise pointed out that the priorities and objectives of nature and landscape conservation are contradictory to their interests and thus incompatible. The Židlochovice Forest Enterprise prefers the production function of the forest that brings them an economic benefit. In contrast, nature conservation authorities try to prevent the loss of forest stands and promote sustainable forest management.

The PLA management plan can be understood as a tool for prescribing restrictions that are not binding on natural persons and legal entities, but it is a binding document for forest holdings and regional planning. Interviews with the representatives of the Moravian Karst PLA and the Pálava PLA reveal a contrast between the participation of local government representatives in the process of drawing up management plans in the localities of interest. While the Pálava PLA administration gives the mayors the final version of the document to look at and any comments are discussed at the Ministry of the Environment, the Moravian Karst PLA administration and the representatives of the municipalities discuss the document at an early stage when the document is being drawn up so the municipality representatives' comments can be considered almost immediately during discussions. The comment of the mayors of municipalities affected by the declaration of the Soutok PLA stems from the insufficient elaboration of the management plan, especially the missing chapters, when the management plan for the proposed Soutok PLA contained only the part dealing with nature protection and human activities affecting the state of nature and landscape. However, it is not common practice to draw up these documents before the PLA is declared.

The most frequent argument against the declaration of the proposed Soutok PLA was that the current protection of the site in question is adequate and the existing care in the form of small-sized specially protected areas and the Lower Morava Biosphere Reserve is sufficient. The basic problem of this argumentation is that small-sized specially protected areas cover only a small part of the proposed site (about 0.94%). Furthermore, the Lower Morava Biosphere Reserve, as the only one of the biosphere reserves in the Czech Republic, is not subject to nature conservation authorities since it is largely situated outside of the large-sized specially protected area (with the exception of the part extending to the Pálava PLA), and it is also the only one represented by a nongovernmental organization. This model of governance is unique because, until now, the management of Czech biosphere reserves has always been associated with the performance of the administration of one of the bodies of the Ministry of the Environment (either a protected landscape area or a national park). The founders are the Czech Union for Nature Conservation, Forests of the Czech Republic, s.p. (state enterprise), MND, a.s. (Moravian Oil Mines), the Ministry of the Environment, and the District Chamber of Commerce Břeclav. Except for the Ministry of the Environment, all the founding bodies were opposed to the declaration of the Soutok PLA. In the meantime, the existing environmental protection allows promoting objectives that damage the environment, which means that it cannot be described as sufficient. Forest management, which is currently incompatible with nature and landscape protection (e.g., wide areas of regeneration blocks, using of exotic productive species game management with a long-time history of planting non-native walnut *Juglans nigra* L.), can also be classified as a fundamental problem that the existing protection of the site does not deal with. Another major change was the regulation of rivers (including the construction of the Nové Mlýny dams, built in the 1970s and 1980s), which meant a significant reduction in the length of their streams and strangling meanders. It had an impact on the entire floodplain forest ecosystem, which now needs to be artificially flooded. The problem also arises of the possible migration of animals to be protected. Managing such a vast territory by way of contractual agreements would only lead to fragmentation of the territory, ineffective management of protection, and an increase in administrative steps.

3.6. Public Administration and Self-Government

The problem of providing incomplete and unilateral information can be described as information asymmetry. The question of communication with the mayors and entities in the affected areas permeated the entire interview.

When efforts were made to declare the Soutok PLA, the opponents also referred to the poor reputation of nature conservation authorities. In their words, the breakthrough moment was the declaration of the Natura 2000 network, which was announced without any previous communication with local authorities. At the same time, the opinions of the mayors whose municipalities belong to the Pálava PLA were also influenced by this fact. Nature and landscape conservation cannot be taken as a strictly prescribed fact, but the general public must be involved in this process too. Residents' foreknowledge and participation in decision making on events in municipalities and their immediate surroundings is an essential part of a democratic society based on the coordination of actions and satisfaction of the entities concerned. It is therefore essential that citizens themselves get involved and submit their proposals and comments. After all, they form an inseparable part of protected areas, and they know best what pitfalls they bring and what problems need to be solved.

Mayors of municipalities affected by the intention to declare the Soutok PLA often stated that PLA brings complication in exercising state administration and self-government. The mayors of municipalities do not find any restrictions caused by the existence of PLA that would limit them in the performance of administrative duties. The situation is similar in the Pálava PLA where the mayors see complications limited to a minimum, and only in the area of construction activity. Municipalities are obliged to include the conditions for future construction activity, which are based on the requirements for nature and landscape conservation, in the master plan documentation. The restrictions are more significant in municipalities with a big part of a cadastral territory or built-up area in PLA. In the case of the Soutok PLA, there is no need for these restrictions since it is defined and delineated outside of the area suitable for construction activity. The mayors also stated that they had no competencies other than the mayors of municipalities outside of the PLA.

The position of local governments should not change even after the PLA is declared. According to the Ministry of the Environment, the only fundamental change will be that the nature conservation authority with competence for the area lying directly in the PLA will be the PLA administration and not the municipality with extended competence or the regional authority. Compared to the current situation when residents need part of the statement from a municipality with extended competence and part from a regional authority, this represents a considerable relief.

4. Discussion

The declaration of protected areas is a frequently discussed and controversial topic since the present society considers nature conservation to be a very restrictive instrument [49]. The general opinion of the society is that conservationists considerably restrict or directly prohibit the alternative use of sites regardless of the changing reality. In some cases, the objectives of the protection are only possible in small scattered locations connected by narrow corridors. According to Chamblee et al. [50], spatial fragmentation of protected areas can bring fewer benefits than the preservation of a large area. In their research, McDonald et al. [51] also confirm the more traditional view that larger areas are more supportive of the protection of sites and enable comprehensive management. It can be argued that in the case of the proposed Soutok PLA protection in the form of large-sized protected areas is preferable. This claim is primarily based on the fact that the proposed PLA should protect the largest complex of floodplain forests in the Czech Republic covering a total area of 139 km^2.

According to Dwyer et al. [52], tourism activities in the region are demonstrated by economic benefits when income and investments are transferred from richer and more developed areas to poorer and less developed areas. The overall aim of "environmentally friendly" tourism should be to achieve a situation in which the benefits generated significantly exceed the losses incurred as a result of its existence [53]. An increase in seasonal jobs can be expected in many municipalities, especially in accommodation and catering facilities. This conclusion is supported by Hall [54] or Kadiyali and Kosová [55], who identify protected areas as an important factor in the regional development of the municipalities concerned. When comparing alternative land use, it is impossible to say precisely which of the options is more economically advantageous for the region since the value of nature is immeasurable. However, to express its value, there is, for example, the concept of ecosystem services, which treats ecosystem functions as services [56] that provide a number of benefits for human well-being [49]. Many of these services can then be expressed by the economic value that is derived from the cost of artificial provisions of these services [56,57]. Although these services are increasingly considered by policymakers in the field of environmental management [58], they are not considered by the general public when deciding on land management [49]. Recognition of ecosystem services as assets of economic and social value can contribute to promoting nature conservation and making more responsible decisions when addressing development projects [59]. Nevertheless, by comparing the resulting value it can be concluded that sustainable management in a given ecosystem is more economically advantageous than its transformation motivated by unilateral land use [45]. Mainly American studies deal with the development of land prices immediately after the declaration of the protected landscape area [50,60]. These studies concluded that the price of land in the immediate vicinity of protected areas tends to rise. However, studies conducted in the United States cannot be considered entirely relevant for the central European region since the land price reflects a number of factors, and a different valuation system may operate on another continent.

The process of establishing and administering a protected area must involve the society, in particular local residents, whose perception of the protected area influences the attitude, actions, and interaction of the residents with the area. Lack of knowledge and attitudes of the main stakeholders can lead to incorrect selection of methods for territory management [25,61,62] and influencing the views of the general public by important players [25]. Stakeholders are thus a key factor for successful site management [23,63]. It is known that a positive attitude toward the protected area has a strong influence on its successful management. However, stakeholders' attitudes and perceptions are not implemented by nature conservation authorities in the decision-making process [64]. Primarily, local stakeholders' socioeconomic factors (age, gender, education, ethnic origin, place of residence, etc.) are examined since they influence their perception of and attitude to protected areas. Only a small part of the study is devoted to their perception of the protected area before its declaration [10,65]. In line with changing societal preferences, environmental consideration has been increasing in forestry in most of the developed world [66]. This has naturally caused conflicts between the traditional utilitarian and environmental interests, e.g., as expressed through the ideological clashes between the defenders of forest owners' rights and the advocates of weightier public interference [66].

Findings from Ward et al. [67] study have highlighted a number of challenges related to PA comanagement as follows: (1) any reduction in ES access is likely to create a short-term opportunity cost. These costs need to be explicitly recognized and livelihood interventions should be designed with this in mind; (2) the diversity of cultural and social values given to livelihood activities relating to ES use needs to be carefully incorporated rather than considering them as conservation or sustainability issues; and (3) community-level PA institutions need to ensure that all household types and social divisions are represented in order to prevent worsening existing or creating new inequalities [67]. Accepting these findings will be crucial for the successful establishment of the Soutok Protected Landscape Area, its acceptance by local stakeholders, and the use of ecosystem services for their quality of life.

5. Conclusions

The declaration of large-sized protected areas in the Czech Republic mostly takes the form of a top-down approach, and thus, it often leads to controversial and emotional debates among politicians, nature conservationists, residents, and stakeholders. Sometimes, the negotiations are held so long that the intention to declare a protected area is abandoned in the end. For example, this was the case of the proposed Soutok PLA. However, it should be noted that the existence of large-sized protected areas is not a barrier to all regional development, and it is even considered to be an engine of the local economy in some countries. Still, it is necessary to differentiate suitable management methods and activities that ensure the development of the area while maintaining its natural and landscape values according to the zones of graded protection. It is essential that socioeconomic development goes hand in hand with successful landscape care and protection. Be it forestry or agricultural activity, there are certain limits that have to be respected. Farmers can claim financial compensation for the loss sustained, which is paid to them but not always in full. Forest management is closely linked with the forest management plan, for which the PLA management plan is binding and must be governed by it. Therefore, mining requirements must be considered; otherwise, a positive opinion will not be issued. In forest management, this should not be a short-term economic benefit from the use of forests. The widespread use of natural processes in forestry practice has a profound and long-term biological effect on the prosperity of forest ecosystems and hence the economy of their management.

The research highlighted the importance of communication between municipality representatives, stakeholders, and residents, on the one hand, and representatives of protected areas, on the other. From the results, it is obvious that communication is the basis for successful cooperation because mutual coordination of steps is necessary for the smooth functioning and coexistence of nature conservation and regional development. Unfortunately, common practice shows that nature conservation authorities often act in a superior manner and do not give local government representatives a choice. This creates a generally negative image of nature conservation. Perfect foreknowledge of the entities on both sides is essential, both for the functioning of nature conservation and for mutual trust necessary for good cooperation.

The research refuted or did not confirm most of the arguments of stakeholders against the declaration of the PLA. Reasoned comments either stemmed from other legislation or did not have a negative impact on regional or municipal development. The approach of nature protection officials toward municipalities and communication with them is perceived as a very important aspect. The negative perception of nature protection and state administration is a serious consequence of a possible unfriendly approach.

The general main problem to be solved is environmental awareness and education, which, for representatives of farmers and owners and municipalities, is especially about the natural benefits, the intrinsic value of nature, and the vulnerability of ecosystems; for nature conservation representatives, this is especially in the area of communication with other stakeholder groups. Both of these tools need to be used for a long time, but also immediately.

Author Contributions: Conceptualization, J.S., A.R. and M.B.; methodology, J.S., A.R. and M.B.; validation, J.S. and A.R.; formal analysis, A.R.; investigation, M.B.; resources, M.B.; data curation, A.R.; writing—original draft preparation, M.B. and A.R.; writing—review and editing, J.S. and A.R.; visualization, M.B. and A.R.; supervision, J.S.; project administration, J.S.; funding acquisition, J.S. All authors have read and agreed to the published version of the manuscript.

Funding: The research was funded by the Internal Grant Agency Faculty of regional development and international studies, Mendel University in Brno, project No. 16/2015—Selected environmental tools as regional development factors and EEA Grants, Grant Number EHP-CZ02-OV-1-032-2015 "Raising awareness and publicity of the importance of forest functions in the landscape and near-natural watercourses in urban areas as a part of basin ecosystem services."

Institutional Review Board Statement: Not applicable.

Informed Consent Statement: Not applicable.

Conflicts of Interest: The authors declare no conflict of interest.

References

1. Cardillo, M.; Mace, G.M.; Gittleman, J.L.; Purvis, A. Latent extinction risk and the future battlegrounds of mammal conser-vation. *Proc. Natl. Acad. Sci. USA* **2006**, *103*, 4157–4161. [CrossRef] [PubMed]
2. Leroux, S.J.; Krawchuk, M.A.; Schmiegelow, F.; Cumming, S.G.; Lisgo, K.; Anderson, L.G.; Petkova, M. Global protected areas and IUCN designations: Do the categories match the conditions? *Biol. Conserv.* **2010**, *143*, 609–616. [CrossRef]
3. Stolton, S.; Dudley, N. *Arguments for Protected Areas: Multiple Benefits for Conservation and Use*; Earthscan: London, UK, 2010; p. 273.
4. Jones, N.; McGinlay, J.; Dimitrakopoulos, P.G. Improving social impact assessment of protected areas: A review of the literature and directions for future research. *Environ. Impact Assess. Rev.* **2017**, *64*, 1–7. [CrossRef]
5. Blondet, M.; De Koning, J.; Borrass, L.; Ferranti, F.; Geitzenauer, M.; Weiss, G.; Turnhout, E.; Winkel, G. Participation in the implementation of Natura 2000: A comparative study of six EU member states. *Land Use Policy* **2017**, *66*, 346–355. [CrossRef]
6. Karanth, K.K.; Naughton-Treves, L.; DeFries, R.; Gopalaswamy, A.M. Living with Wildlife and Mitigating Conflicts Around Three Indian Protected Areas. *Environ. Manag.* **2013**, *52*, 1320–1332. [CrossRef]
7. Rakotonarivo, O.S.; Jacobsen, J.B.; Poudyal, M.; Rasoamanana, A.; Hockley, N. Estimating welfare impacts where property rights are contested: Methodological and policy implications. *Land Use Policy* **2018**, *70*, 71–83. [CrossRef]
8. Zorondo-Rodríguez, F.; Díaz, M.; Simonetti-Grez, G.; Simonetti, J.A. Why would new protected areas be accepted or rejected by the public?: Lessons from an ex-ante evaluation of the new Patagonia Park Network in Chile. *Land Use Policy* **2019**, *89*, 104248. [CrossRef]
9. Barbić, A.; Udovč, A.; Medved, A. Protection of the environment and biodiversity for sustainable future of rural areas: The case of planned Regional park Trnovski gozd, Slovenia. *Sociol. Sela* **2004**, *42*, 4.
10. Winter, S.J.; Esler, K.J.; Kidd, M. An index to measure the conservation attitudes of landowners towards Overbeg Coastal Renosterveld, a critically endangered vegetation type in the Cape Floral Kingdom, South Africa. *Biol. Conserv.* **2005**, *126*, 383–394. [CrossRef]
11. Nastran, M. Why does nobody ask us? Impacts on local perception of a protected area in designation, Slovenia. *Land Use Policy* **2015**, *46*, 38–49. [CrossRef]
12. Conrad, E.; Fazey, I.; Christie, M.; Galdies, C. Choosing landscapes for protection: Comparing expert and public views in Gozo, Malta. *Landsc. Urban Plan.* **2019**, *191*, 103621. [CrossRef]
13. Allendorf, T.D. A framework for the park–people relationship: Insights from protected areas in Nepal and Myanmar. *Int. J. Sustain. Dev. World Ecol.* **2010**, *17*, 417–422. [CrossRef]
14. MacKenzie, C.A.; Salerno, J.; Hartter, J.; Chapman, C.A.; Reyna, R.; Tumusiime, D.M.; Drake, M. Changing perceptions of protected area benefits and problems around Kibale National Park, Uganda. *J. Environ. Manag.* **2017**, *200*, 217–228. [CrossRef]
15. Pullin, A.S.; Bangpan, M.; Dalrymple, S.E.; Dickson, K.; Haddaway, N.R.; Healey, J.R.; Hauari, H.; Hockley, N.; Jones, J.P.G.; Knight, T.M.; et al. Human well-being impacts of terrestrial protected areas. *Environ. Évid.* **2013**, *2*, 19. [CrossRef]
16. de Lange, E.; Woodhouse, E.; Milner-Gulland, E.J. Approaches used to evaluate the social impacts of protected areas. *Conserv. Lett.* **2016**, *9*, 327–333. [CrossRef]

17. Kaplan-Hallam, M.; Bennett, N.J. Adaptive social impact management for conservation and environmental management. *Conserv. Biol.* **2018**, *32*, 304–314. [CrossRef]
18. Schmitz, M.; Matos, D.; de Aranzabal, I.; Ruiz-Labourdette, D.; Pineda, F. Effects of a protected area on land-use dynamics and socioeconomic development of local populations. *Biol. Conserv.* **2012**, *149*, 122–135. [CrossRef]
19. Bennett, N.J. Using perceptions as evidence to improve conservation and environmental management. *Conserv. Biol.* **2016**, *30*, 582–592. [CrossRef]
20. Abukari, H.; Mwalyosi, R.B. Local communities' perceptions about the impact of protected areas on livelihoods and community development. *Glob. Ecol. Conserv.* **2020**, *22*, e00909. [CrossRef]
21. Ciocănea, C.M.; Sorescu, C.; Ianoşi, M.; Bagrinovschi, V. Assessing Public Perception on Protected Areas in Iron Gates Natural Park. *Procedia Environ. Sci.* **2016**, *32*, 70–79. [CrossRef]
22. Buta, N.; Holland, S.M.; Kaplanidou, K. Local communities and protected areas: The mediating role of place attachment for pro-environmental civic engagement. *J. Outdoor Recreat. Tour.* **2014**, *5*, 1–10. [CrossRef]
23. Arnberger, A.; Eder, R.; Allex, B.; Sterl, P.; Burns, R.C. Relationships between national-park affinity and attitudes towards protected area management of visitors to the Gesaeuse National Park, Austria. *For. Policy Econ.* **2012**, *19*, 48–55. [CrossRef]
24. Trakolis, D. Local people's perceptions of planning and management issues in Prespes Lakes National Park, Greece. *J. Environ. Manag.* **2001**, *61*, 227–241. [CrossRef]
25. Allendorf, T.D.; Smith, J.L.; Anderson, D.H. Residents' perceptions of Royal Bardia National Park, Nepal. *Landsc. Urban Plan.* **2007**, *82*, 33–40. [CrossRef]
26. Spiteri, A.; Nepalz, S.K. Incentive-Based Conservation Programs in Developing Countries: A Review of Some Key Issues and Suggestions for Improvements. *Environ. Manag.* **2005**, *37*, 1–14. [CrossRef]
27. Apostolopoulou, E.; Drakou, E.G.; Pediaditi, K. Participation in the management of Greek Natura 2000 sites: Evidence from a cross-level analysis. *J. Environ. Manag.* **2012**, *113*, 308–318. [CrossRef] [PubMed]
28. Rojas-Briales, E. Socio-economics of nature protection policies in the perspective of the implementation of Natura 2000 Network: The Spanish case. *Forestry* **2000**, *73*, 199–207. [CrossRef]
29. Nastran, M.; Pirnat, J. Stakeholder Participation in Planning of the Protected Natural Areas: Slovenia. *Sociol. Prost.* **2012**, *19350*, 141–164. [CrossRef]
30. Schneider, J.; Ruda, A.; Kalasová, Ž.; Paletto, A. The Forest Stakeholders' Perception towards the NATURA 2000 Network in the Czech Republic. *Forests* **2020**, *11*, 491. [CrossRef]
31. Vyskot, I.; Kupec, P.; Schneider, J.; Klimánek, M. Optimalisation of forest management–model area of Special Protection Area of Soutok–Tvrdonicko, Forest Enterprise Zidlochovice, Forests of the Czech Republic. In *Biodiversity and Target Management of Floodplain Forests in Morava River Basin*, 1st ed.; Machar, I., Ed.; Nakladatelství Univerzity Palackého v Olomouci: Olomouc, Czech Republic, 2010; pp. 172–181. ISBN 978-80-244-2530-6.
32. Mackovčin, P.; Jatiová, M.; Demek, J.; Slavík, P. Brněnsko. In *Chráněná Území ČR, Svazek IX*; Ekocentrum Brno a VÚKOZ; Mackovčin, P., Sedláček, M., Eds.; AOPK ČR: Praha, Czech Republic, 2007.
33. PLA Moravský Kras. *Management Plan for PLA Moravský Kras for Period 2019–2028*; PLA Moravský Kras: Brno, Czech Republic, 2017.
34. Bárta, F.; Němec, J.; Pojer, F. *Krajina v České Republice*; Consult: Praha, Czech Republic, 2007; p. 399.
35. PLA Pálava. Management Plan for PLA Pálava for Period 2016–2025. 2015. Available online: https://palava.ochranaprirody.cz/ochrana-prirody/plan-pece-chko-palava/ (accessed on 26 July 2019).
36. Mayring, P. Qualitative content analysis. In *A Companion to Qualitative Research*; Flick, U., von Kardoff, E., Steinke, I., Eds.; Sage: London, UK, 2004; pp. 266–269.
37. Mayring, P. *Qualitative Inhaltsanalyse. Grundlagen und Techniken*; Beltz: Weinheim, Germany, 1983.
38. Mayring, P. Qualitative Content Analysis. Forum Qualitative Sozialforschung; Forum: Qualitative Social Research 1. 2000. Available online: https://www.qualitative-research.net/index.php/fqs/article/view/1089b (accessed on 26 July 2019).
39. Bazeley, P. *Qualitative Data Analysis with NVivo*, 2nd ed.; Sage: Thousand Oaks, CA, USA, 2007.
40. Mayring, P. Qualitative Content Analysis: Theoretical Foundation, Basic Procedures and Software Solution. 2014. Available online: https://www.researchgate.net/publication/266859800_Qualitative_content_analysis_-_theoretical_foundation_basic_procedures_and_software_solution (accessed on 26 July 2019).
41. Richards, T. An intellectual history of NUD*IST and NVivo. *Int. J. Soc. Res. Methodol.* **2002**, *5*, 199–214. [CrossRef]
42. Glaser, B.G.; Strauss, A.L. The Discovery of Grounded Theory: Strategies for Qualitative Research; 1999. Available online: http://www.sxf.uevora.pt/wp-content/uploads/2013/03/Glaser_1967.pdf (accessed on 26 July 2019).
43. Bryant, A.; Charmaz, K. Grounded Theory in Historical Perspective: An Epistemological Account. In *The SAGE Handbook of Grounded Theory*; Bryant, A., Charmaz, K., Eds.; SAGE: London, UK, 2007; pp. 31–57.
44. Act, no. 114/1992 Coll. on Nature Conservation and Landscape Protection. Sbírka zákonů České republiky. 1992.
45. Košulič, M. *Cesta k Přírodě Blízkému Hospodářskému Lesu*; FSC Česká Republika—Forest Stewardship Council: Brno, Czech Republic, 2010; p. 452.
46. Act, no. 338/1992 Coll. on the Tax on Immovables. Sbírka zákonů České republiky. 1992.
47. Act, no. 243/2000 Coll. on Budget Allocation of Taxes. Sbírka zákonů České republiky. 2000.

48. Abelairas-Etxebarria, P.; Astrokiza, I. Farmland prices and land-use changes in periurban protected natural areas. *Land Use Policy* **2012**, *29*, 674–683. [CrossRef]
49. Seják, J. Principy a metody oceňování životního prostředí. *Zivotn. Prostr.* **2002**, *36*, 10–13.
50. Chamblee, J.F.; Colwell, P.F.; Dehring, C.A.; Depken, C.A. The Effect of Conservation Activity on Surrounding Land Prices. *Land Econ.* **2011**, *87*, 453–472. [CrossRef]
51. McDonald, R.I.; Yuan-Farrell, C.; Fievet, C.; Moeller, M.; Kareiva, P.; Foster, D.; Gragson, T.; Kinzig, A.; Kuby, L.; Redman, C. Estimating the Effect of Protected Lands on the Development and Conservation of Their Surroundings. *Conserv. Biol.* **2007**, *21*, 1526–1536. [CrossRef]
52. Dwyer, L.; Forsyth, P.; Dwyer, W. *Tourism Economics and Policy*; Channel View Publications: Bristol, UK, 2010; Volume 5, p. 880.
53. Christofakis, M. Strategic Options for Tourism Impacts on Local Sustainability: A Conceptual Approach. *Local Econ.* **2010**, *25*, 586–598. [CrossRef]
54. Hall, C.M. *Tourism—Rethinking the Social Science of Mobility*; Pearson Prentice Hall: London, UK, 2005.
55. Kadiyali, V.; Kosová, R. Inter-industry employmeent spillovers from tourism inflows. *Reg. Sci. Urban Econ.* **2013**, *43*, 272–281. [CrossRef]
56. Pithart, D.; Křováková, K.; Žaloudík, J.; Dostál, T.; Valentová, J.; Valenta, P.; Weyskrabová, J.; Dušek, J. Ecosystem services of natural floodplain segment—Lužnice River, Czech Republic. In *Flood Recovery, Innovation and Response II*; Wrachien, D., Proverbs, D., Brebbia, C.A., Mambretti, S., Eds.; WIT Press: Southampton, UK, 2010; pp. 129–143.
57. De Groot, R.S.; Alkemade, R.; Braat, L.; Hein, L.; Willemen, L. Challenges in integrating the concept of ecosystem services and values in landscape planning, management and decision making. *Ecol. Complex.* **2010**, *7*, 260–272. [CrossRef]
58. Boithias, L.; Terrado, M.; Corominas, L.; Ziv, G.; Kumar, V.; Marqués, M.; Schuhmacher, M.; Acuña, V. Analysis of the un-certainty in the monetary valuation of ecosystem services—A case study at the river basin scale. *Sci. Total Environ.* **2016**, *543*, 683–690. [CrossRef]
59. Fei, L.; Shuwen, Z.; Jiuchun, Y.; Kun, B.; Qing, W.; Junmei, T.; Liping, C. The effects of population density changes on ecosystem services value: A case study in Western Jilin, China. *Ecol. Indic.* **2016**, *61*, 328–337. [CrossRef]
60. Geoghegan, J. The value of open spaces in residental land use. *Land Use Policy* **2002**, *19*, 91–97. [CrossRef]
61. Reed, M.S. Stakeholder participation for environmental management: A literature review. *Biol. Conserv.* **2008**, *141*, 2417–2431. [CrossRef]
62. King, B.; Peralvo, M. Coupling Community Heterogeneity and Perceptions of Conservation in Rural South Africa. *Hum. Ecol.* **2010**, *38*, 265–281. [CrossRef]
63. Arnberger, A.; Schoissengeier, R. The other side of the border: Austrian local residents' attitudes towards the neighbouring Czech Šumava National Park. *J. Nat. Conserv.* **2012**, *20*, 135–143. [CrossRef]
64. Struhsaker, T.T.; Struhsaker, P.J.; Siex, K.S. Conserving Africa's rain forests: Problems in protected areas and possible solutions. *Biol. Conserv.* **2005**, *123*, 45–54. [CrossRef]
65. Barbić, A.; Udovč, A. *Protection of the Environment and Biodiversity for Sustainable Future of Rural Areas: The Case of Planned Regional Park Trnovski Gozd, Slovenia*; Discussion Papers 18886; CEESA—Central and Eastern European Sustainable Agriculture International Research Project: Trnovski Gozd, Slovenia, 2003; Available online: http://hdl.handle.net/10419/22296 (accessed on 26 July 2019).
66. Brukas, V.; Stanislovaitis, A.; Kavaliauskas, M.; Gaižutis, A. Protecting or destructing? Local perceptions of environmental consideration in Lithuanian forestry. *Land Use Policy* **2018**, *79*, 1014–1023. [CrossRef]
67. Ward, C.; Stringer, L.; Holmes, G. Changing governance, changing inequalities: Protected area co-management and access to forest ecosystem services: A Madagascar case study. *Ecosyst. Serv.* **2018**, *30*, 137–148. [CrossRef]

forests

MDPI

Article

Efficient, Sustainable, and Multifunctional Carbon Offsetting to Boost Forest Management: A Comparative Case Study

Timothée Fouqueray [1,*], Lucile Génin [1], Michel Trommetter [2] and Nathalie Frascaria-Lacoste [1]

[1] Ecologie Systématique Evolution, CNRS, AgroParisTech, Université Paris-Saclay, 91405 Orsay, France; lucile.genin@sciencespo.fr (L.G.); nathalie.frascaria@universite-paris-saclay.fr (N.F.-L.)

[2] GAEL, Grenoble INP, CNRS, INRAE, Université Grenoble Alpes, 38000 Grenoble, France; michel.trommetter@inrae.fr

* Correspondence: timothee.fouqueray@uqo.ca

Abstract: Research highlights: Funding forest management with subsidies from carbon offsetters is a well-documented mechanism in tropical regions. This article provides complementary insights into the use of voluntary offset contracts in temperate forests. Background and objectives: The mitigation of greenhouse emissions has become a major global issue, leading to changes in forest management to increase the capacity of forests to store carbon. This can lead to conflicts of use with other forest ecosystem services such as timber production or biodiversity conservation. Our main goal is to describe collective actions to fund carbon-oriented forestry with subsidies from carbon offsetters and to analyze how their governance and functioning prevent conflicts pertaining to multi-functionality. Materials and methods: We assembled an interdisciplinary research team comprising two ecologists, a social scientist, and an economist. Drawing on a conceptual framework of ecosystem services, social interdependencies, and collective action, we based our qualitative analysis on semi-structured interviews from two French case studies. Results: Carbon-oriented intermediary forest organizations offer offset contracts to private firms and public bodies. Communication is geared toward the mitigation outcomes of the contracts as well as their beneficial side effects in providing the ecosystem services of interest to the offsetters. Subsidies then act as a financial lever to fund carbon-oriented forestry operations. Scientific committees and reporting methodologies serve as environmental, social, and economic safeguards. Conclusions: These new intermediary forest organizations use efficient forest operations and evaluation methodologies to improve forest carbon storage. Their main innovation lies in their collective governance rooted in regional forest social-ecological systems. Their consideration of multi-functionality and socioeconomic issues can be seen as an obstacle to rapid development, but they ensure sustainability and avoid conflicts between producers and beneficiaries of forest ecosystem services. Attention must be paid to interactions with broader spatial and temporal carbon policies.

Keywords: forest management; multifunctionality; carbon offset; collective action; conflict avoidance; mitigation; payment for ecosystem services; social-ecological; France

Citation: Fouqueray, T.; Génin, L.; Trommetter, M.; Frascaria-Lacoste, N. Efficient, Sustainable, and Multifunctional Carbon Offsetting to Boost Forest Management: A Comparative Case Study. *Forests* **2021**, *12*, 386. https://doi.org/10.3390/f12040386

Academic Editor: Jessica Leahy

Received: 22 February 2021
Accepted: 22 March 2021
Published: 24 March 2021

1. Introduction

The claim that forests and foresters have witnessed significant evolutions in the last few decades is an understatement. Along with social and technical developments, it is now routine to point to global changes as drivers of the spatiotemporal trajectories of forest socio-ecosystems (FSES) [1]. Among the most pressing global issues (with land degradation), climate change has buoyed considerable research efforts in the forest sciences [2,3].

While undeniable research progress has been made for the mitigation of and adaptation to climate change, the indirect consequences of mitigation and adaptation actions on FSES remain underexplored. To cite only mitigation, a sizeable body of literature explores the direct influence of forest management on above- and below-ground carbon storage [4]. However, mitigation is rarely tackled from the perspective of its social or economic consequences (e.g., [5])—with the notable exceptions of investigations into the Reducing Emissions from Deforestation and Forest Degradation (REDD+) program [6] and in some European case studies [7,8].

Although it makes sense to monitor how carbon-oriented operations directly influence the role of forests as carbon sinks, the neutral, positive, or negative effects of such operations on forest multifunctionality should not be overlooked [8]. Indeed, many forest stakeholders deem the simultaneous provision of different forest ecosystem services (ES) to be critical [9,10], especially the provision of timber and non-timber forest products, but also the esthetic value of managed landscapes for local communities or tourism development and the role of forest soils in water epuration, to name but a few [11].

This consideration of multifunctionality in the management of FSES also aligns with public opinion on carbon storage in forests. Particularly in Western countries, consumer-citizens are putting mounting pressure on public institutions and the private sector to develop carbon offsetting schemes that do not neglect biodiversity conservation or local employment issues [12]. Turning to forest management thus gives carbon offsetters the opportunity to "kill three birds with one stone." First, forests represent an efficient and powerful lever to mitigate climate change—they already represent 14.5% of the total volume of greenhouse gas (GHG) emissions saved on voluntary markets since 2005 [13]. Second, forests benefit from a highly natural perception in civil society [14]. For companies assigning shares to sustainable forest management, this eases their response to the rising social demands for businesses to assume part of both the fight against global warming and the biodiversity crisis. Third, the frequent relocation of offset projects from tropical to temperate and boreal forests allows companies to align their environmental and social commitments [15]. Indeed, such relocations often occur in places with sound property rights, thus favoring local governance and employment, and avoiding the pitfalls of remote and potentially illegible operations, as observed with the REDD+ programs [6,16].

With its forest and mitigation policies, France is a suitable illustration of this renewed interest in local forests from carbon offsetters. In terms of forests, France has Europe's fourth largest surface (165,000 km^2), most of which is deciduous [17]. In managed forests, a tradition of multifunctionality has been legally embedded into policies, with pledges for the simultaneous provision of various ES such as timber production, recreational activities, and biodiversity conservation [18]. In terms of voluntary mitigation, France organized cross-sectional meetings in 2007 with businesses, public collectivities, and civil society stakeholders in order to promote GHG mitigation [19]. Several recommendations were later implemented in a law on corporate social and environmental responsibility (CSR), including mandatory carbon assessments [19,20]. Therefore, it is unsurprising that many voluntary mitigation projects have blossomed in the French forest sector since the early 2010s [21,22]. These projects are based on unprecedented partnerships linking forest owners, forestry experts, and timber industry professionals with entities seeking to compensate their GHG emissions.

The rationale of this paper is to investigate how voluntary mitigation projects boost forest management from the perspective of three essential issues, namely the efficiency, sustainability, and multifunctionality of carbon storage. Drawing deductively on a literature review on forest carbon storage [23] and inductively on our experience with forest practitioners [9], we considered that these were the three best factors predicting the success of mitigation projects. The first factor relates to the efficiency of carbon offsetting. Following Gren and Aklilu [23], we posit that any offset project must consider heterogeneity, uncertainty, additionality, and permanence issues in order to be efficient. The most important issue is proving the added value of the project (additionality), meaning that the forestry operations would not have occurred without the funding of the contract [24,25]. Second, the permanence risk refers to the potential carbon leakage during the duration of the offset project [15]. Third, the uncertainty of forest growth makes it difficult to predict the mitigation outcome of forestry operations [26]. Finally, the heterogeneous capacities of forests to store carbon complicates the calculation of the anticipated mitigation success. The second factor relates to the sustainability of carbon offsetting. To endure in the long term and recruit carbon offsetters, voluntary carbon offsetting projects must prove their added value (or "economic additionality") in comparison to baseline forest operations. The sustainability of the offset projects also implies the prevention of social conflict. Even if such conflict is not desired by FSES stakeholders, they should at least prevent any opposition. The third factor follows from the second: we assume that attention must be paid to respecting FSES multifunctionality in order to achieve carbon offsetting. Indeed, multi-objective forest operations are often inconsistent with a management geared toward only one ES, a situation commonly illustrated through the intensification of management for timber production (e.g., shortened rotation length), which can go against biodiversity conservation or the preservation of water quality among others [27–29].

Section 2 presents comparative case studies of two French organizations that bring together forest stakeholders and carbon offsetters from the private and public sectors. It also includes the conceptual framework of Barnaud et al. on ES, social interdependencies, and collective action [30] on which we based our analysis, and outlines any methodological considerations. Section 3 describes the results and depicts how the shared objective of carbon offsetting raises various organizational and governance issues. Section 4 summarizes the results of this national French study and draws conclusions at a broader scale.

2. Materials and Methods

2.1. Selection of Case Studies

The selection of case studies was guided by two steps to identify unique partnerships linking forest owners, forestry experts, and timber industry professionals (hereafter, foresters) to carbon offsetters. We restricted the area to France to allow for comparisons within the same legal and institutional framework. First, we conducted a Google survey to list French carbon mitigation projects linked to forests (Appendix A Table A1) using all possible combinations of the French keywords "forêt," "compensation carbone," and "climat" (signifying "forest," "carbon offsetting," and "climate," respectively). We crosschecked the list with a record of conferences held on forest-based carbon mitigation. The record dating back to 2016 was established using newsletters from forestry or environmental websites (https://www.actu-environnement.com, https://www.fransylva.fr (accessed on 5 October 2018), https://www.academie-agriculture.fr (accessed on 5 October 2018), https://www.alternativesforestieres.org (accessed on 5 October 2018)), and using the mapping of carbon projects available at https://www.cnpf.fr/n/nos-partenariats-carbone/n:2493 (accessed on 5 October 2018)). Second, the selection of study areas was guided by social-ecological considerations so as to retain two local initiatives that were similar in scope, partnership structure, and objectives. We did not aspire to cover the entire range of French forested regions or carbon offset organizations.

We chose two nonprofit organizations from two different regions: Sylv'ACCTES founded in 2015 and Normandie Forêver (hereafter, NF) founded in 2017. The study area of Sylv'ACCTES is in the Auvergne-Rhône-Alpes region, which has a long history of fuelwood exploitation. As one of the regions with the highest standing volume, it is dominated by a mixture of coniferous species in the mountains and broadleaved beech (*Fagus sylvatica* L.) and chestnut (*Castanea sativa* Mill.) in the plains [31]. By contrast, the second study area in Normandy is a coastal hilly area with a few forested patches of broadleaved beech and oak (*Quercus robur* L. and *Quercus petraea* (Matt.) Liebl.).

Throughout the article, interviewees are denoted by their anonymization number preceded by "S" or "NF" for Sylv'ACCTES and Normandie Forêver, respectively (for an overview, see Appendix B Table A2).

2.2. Data Collection

We carried out 14 face-to-face, semi-structured, open-ended interviews in November and December 2018 (Appendix B Table A2). We aimed to have a representative and proportional set of interviews for the two nonprofit organizations, balanced by their regional weight and level of activity (11 interviews took place in Auvergne-Rhône-Alpes and 3 in Normandy). Interviewees included forest managers, scientific committee members, financial partners such as public bodies and state agency officials, and forestry experts. Each interview followed a similar grid (Appendix B) consisting of the following: (i) a general description of the interviewee's role, objectives, and experience in the organization; (ii) the interviewee's view about the organization's ability to create bonds between stakeholders; and (iii) the interviewee's vision about the impact and potential development of the organization in the future. If not spontaneously mentioned by the interviewees, we also asked them about the implementation of the official "low carbon certification" (LCC) recently launched by the French Ministry of the Environment and its effect on the organization.

We recorded and fully transcribed the interviews, which lasted between 1 and 2.5 h. We supplemented them with a review of the forestry press with a focus on mitigation projects (the newsletters are listed in Section 2.1), the websites of Sylv'ACCTES and NF (https://sylvacctes.org and http://www.normandieforever.org, both accessed on 5 October 2018), as well as their preliminary reports on carbon accounting, operational charters, partnership leaflets, or financial information. In the following, any results without an interviewee number are sourced from this complementary documentation.

Our analysis aimed to understand the content of the offset contracts and their framework, namely the regional constraints of NF and Sylv'ACCTES, how they developed their organizational processes (distribution of decisional power between stakeholders, jurisdictional form, spatial and temporal scales, financing, etc.), and the importance given to carbon mitigation by the contractors (e.g., methodological soundness accounting for GHG mitigation). We thus developed a qualitative analysis grid (Appendix C to tag and evaluate the transcription of each interview depending on the above items. After its completion by one author, the characterizations were crosschecked by a second author to avoid arbitrary classifications.

2.3. *Data Analysis: Methodological Approach*

We considered that the conceptual framework of Barnaud et al. on ES, social interdependencies, and collective action [30] was the most appropriate to investigate the efficiency, sustainability, and multifunctionality of NF and Sylv'ACCTES. Indeed, ES are at the core of this framework, which stresses how ES providers, beneficiaries, and intermediaries evolve in an "action arena" made of synergetic, antagonistic, or neutral interdependencies. The interdependencies can relate to different ES providers (e.g., private landholders, public forest managers), ES providers and beneficiaries (e.g., forest managers whose stand regeneration suffers from the overgrazing of boars and deer, hunters who are responsible for preventing the overpopulation of game animals), and ES beneficiaries (e.g., motorcyclists, hikers). These interdependencies are reshaped by changes in management decisions that modify ES and by new trade-offs in the interests of forest stakeholders. Intermediaries promoting collective actions such as NF and Sylv'ACCTES also contribute to the permanent restructuring of these social relationships. In the case of carbon offsetting, the analysis of the action arena should embrace four key dimensions: (i) stakeholders' representation of the FSES in light of carbon storage (i.e., "cognitive framing"); (ii) the spatial and temporal levels of organization at which carbon storage occurs; (iii) existing formal and informal institutions regulating carbon storage; and (iv) power relations between FSES stakeholders.

Rooted in Ostrom's work, this social-ecological framework accords with the conflict-free situations of NF and Sylv'ACCTES, in which carbon storage is seen as a public good and not as a product to be sold on carbon markets. To best apply this interdisciplinary conceptual framework, our research team was comprised of two ecologists, a social scientist, and an economist. This framework operates best in situations without open conflict where the emerging organization stems from a collective action around a public or common good (i.e., "a voluntary process of cooperation among various stakeholders, users, and managers addressing a common ES management problem in a given territory" [30]). We, therefore, discarded other conceptual frameworks such as the theories of non-state market-driven governance [32].

3. Results

To facilitate the comprehension of this section, Figure 1 summarizes the general functioning of the two nonprofit organizations: voluntary candidates, either private companies or public bodies, become members of the carbon offset organization after agreeing to its ethical conditions. They provide funding that is managed by forest practitioners. The organization then contracts carbon offsets with forest landholders.

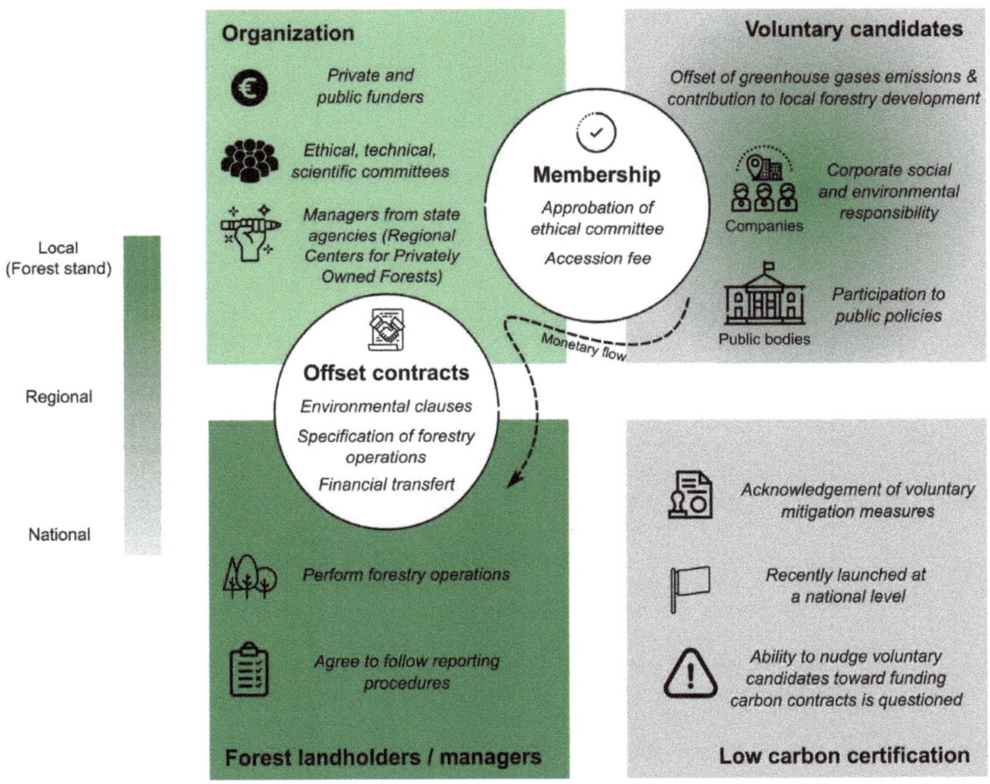

Figure 1. Internal functioning of the organizations offering voluntary carbon offset contracts.

3.1. *Offset Contracts to Increase Carbon Storage*

Offset contracts given by Sylv'ACCTES and NF fund the forestry operations of forest landowners provided that they respect the technical specifications (Figure 1). For instance, afforestation, deforestation, or reforestation are not eligible technical itineraries with Sylv'ACCTES; its contracts instead favor an evolution toward improved forest management (e.g., introduction of species mixtures rather than monospecific stands). Sylv'ACCTES also takes local hazards into account in order to exclude any detrimental effects on the local populations. In a flood-prone area, for instance, the funding of a project was subjected to the conservation of a continuous forest cover. Contrary to Sylv'ACCTES, NF allows a diversity of forest operations (e.g., clear-cut harvesting and replantation), as long as they occur in forest stands in which timber sales do not cover the cost of forestry operations. This difference may be due to the inappropriate planting of tree species given the local conditions or previous parasite attacks decreasing the value of the trees, among others. Sylv'ACCTES excludes contracts for forested surfaces under 2 ha [33].

To gauge how much carbon would be stored in the forest stands included in the offset contracts, the scientific committees of Sylv'ACCTES and NF developed reporting methods: "We have a scientific committee whose job is to make calculations and validate them, but also to confirm or refuse the proposals of forestry sites" (NF1). Prevailing issues around the reporting methods related to the heterogeneity, uncertainty, and impermanence of forests' capacities to store carbon as well as the additionality of the offset contracts.

First, the two organizations need to account for the heterogeneous soil, climate, or tree conditions of the stands. The reporting methodologies usually calculate the amount of carbon saved as the difference in carbon stored between a baseline scenario and a reference scenario. For both organizations, the baseline scenarios (no offset contracts) correspond to the management practices currently in place in the forest stands—for instance, the non-management of decaying stands for NF. Reference scenarios (with offset contracts) are the management of even-aged plantations for NF, while they are chosen among local forestry practices for Sylv'ACCTES (e.g., intensive even-aged forestry) (S1, NF1). Forest experts and ecologists selected practices with an added value for carbon mitigation such as the transition from even-aged forests to species- and age-mixed forests.

Second, the reporting methods need to account for potential carbon mitigation losses due to calculation uncertainties and stochastic natural uncertainties. Sylv'ACCTES retains estimations of carbon storage based on the most optimistic climate scenarios. Once calculated, only 80% of the carbon storage estimation is considered: in this case, if a project suffers from unexpected carbon leakage, it is assumed that the 20% uncounted carbon from all other projects can compensate for the mitigation shortfall. This methodology was patented under the name "Potential Mitigation Gain" (www.sylvacctes.org/les-indicateurs-sylv-acctes/, (accessed on 5 October 2018). NF adopted a similar strategy: to compensate for the potential carbon losses, 15% to 20% of the carbon stored is set aside in every calculation (NF1).

Third, the two carbon organizations aim to avoid carbon leakage during the engagement period of the offset contracts (10 years for Sylv'ACCTES and 20 years for NF). Only NF introduced a replanting clause into the contract in the event of carbon release following a natural hazard (e.g., parasite outbreaks, fires). The impermanence of carbon storage can also stem from contract violations if a forest owner harvests trees before the term of the contract. If the contracted activities are not properly completed, both organizations impose financial penalties—contractors are warned in advance about the possibility of verifications. Sylv'ACCTES dedicates a specific budget to conducting verifications (S1). For financial reasons, NF decided to verify the process internally rather than through external certification: "For us, the certification costs were too high to certify a project on the voluntary mitigation market" (NF3).

The fourth issue of the reporting methodology—additionality—was viewed as very delicate. Two approaches dealing with the issue of non-additionality are found in the literature [23]. The first is to accept the extent of non-additionality in the design of mitigation measures using "non-additionality buffers" among others (hence, additional to the uncertainty buffers). These buffers can be material, with the reduced counting of the carbon stored by the mitigation project, for instance, by setting aside 20% of GHG emissions of an offset project. Otherwise, buffers can be based on an insurance system: in the case of the disclosure of a non-additionality flaw, a saved part of the credits is dedicated to implementing complementary mitigation projects. The second approach used to ensure additionality is to have a contract framework that allows for a case-by-case analysis of the non-additionality risks. Sylv'ACCTES and NF chose the second option, with NF even adding an additionality clause to the offset contracts: "I hereby certify to the best of my knowledge that the funding provided by Normandie Forêver had an active and incentivizing role in my reforestation action for this stand, which had been caught in a forestry deadlock."

Thanks to its offset contracts launched between 2016 and 2019, Sylv'ACCTES has participated in forestry operations over a surface of 4000 ha, split across 20 different forest stands in Rhône-Alpes. Despite its later start in 2018, NF's contracts have already supported forestry operations over 20 ha. NF now aims to reforest 1000 to 2000 ha of forests per year. The main concern of many interviewees was thus not the relevance of the offset contracts but rather the need to recruit enough forest landowners: "If many people wanted to fund carbon offsetting, but there weren't enough owners, then we would be headed for trouble" (NF2). A Sylv'ACCTES' member explained that offset contracts are not given full consideration by private owners and that private forests only represent 20% of grantees, with the remaining 80% used for municipal forests (S1). A NF interviewee also pointed to the overrepresentation of owners aged over 60 years who represent two-thirds of forest owners: "We are among a relatively old group of people, so it is difficult to speak of 20, 30, or 100 years into the future with someone who is already 80" (NF2). Thus, both organizations are in the process of hiring a permanent employee devoted to contacting landholders.

Based on their experience, Sylv'ACCTES and NF members were invited to participate in the creation of the national "low carbon certification." Prior to 2018, the lack of a carbon standard in France led to the search for a reliable tool for voluntary carbon offsetting in order to meet societal expectations regarding GHG mitigation. Although some interviewees acknowledged the value of this initiative, they questioned the technical feasibility of this results-oriented LCC: "The problem with carbon projects is that we must do what is called a proof of additionality to prove that what we do is better than baseline. And in forests, well ... " (S1).

3.2. Carbon Organizations: Pursuing Multifunctional Forestry to Attract Funders

To provide appropriate financing for their offset contracts, Sylv'ACCTES and NF depend on the financial support of both private and public bodies, attracted by communication campaigns focused on the mitigation of GHG emissions. For instance, NF's leaflet includes many such expressions ("A concrete and guaranteed commitment to reducing CO_2 emissions by local reforestation"), while the very name of Sylv'ACCTES stands for "Sylviculture d'Atténuation du Changement Climatique et Services Ecosystémiques" (French for "Silviculture for the Mitigation of Climate Change and Ecosystem Services"). Increasing carbon storage and sequestration in forests was an objective shared by all public and private funders of the associations. Public funders such as towns, cities, and regional councils expressed their interest in forest operations to improve GHG mitigation: "Sylv'ACCTES has another advantage: working with carbon throughout the entire wood lifecycle, and in addition to the societal demands, this is a strong political objective at every level" (S3). In the same vein, the creation of NF was driven by companies wishing to mitigate their carbon emissions: "The sustainable development club—for companies that are very involved in CSR, it is a network. They wondered how they could reduce carbon emissions, and they were looking for local mitigation solutions" (NF1). In Normandy, private funders even prioritize carbon mitigation over other considerations. The aim is to stay as effective as possible and avoid interference from other issues, because when there are "expectations other than carbon such as biodiversity, environment, landscape, or satisfying locally elected officials, it just becomes much more political. By staying focused on carbon, we chose to prioritize it—without removing the rest, we just prioritized carbon" (NF1).

To put this interest into funding, NF relies on two mechanisms that are open to any private or public structure: an annual membership fee of 150 € (Figure 1), and a contribution to NF of 14 € per carbon ton emitted annually by the new member. To avoid greenwashing or conflicts of interest, membership is subject to an ethical code and the decision of the ethical committee: "As firms are highly invested in CSR, they want it to be a real process of carbon mitigation, so we established rules to join Normandie Forêver" (NF1). For Sylv'ACCTES, a membership fee is the first source of income and demonstrates a willingness to contribute to the organization's decisions (Figure 1). Second, the organization benefits from the monetary support of its public and private funding members at a regional level (Auvergne-Rhône-Alpes region, city of Lyon, and Neuflize Bank) and at a smaller scale (local businesses contribute to funding local forestry operations). In addition, Sylv'ACCTES benefits from an asset that NF does not have. Whereas both are nonprofit organizations, only Sylv'ACCTES is recognized to be of general interest by the French state due to its commitment "to the defense of the natural environment" (Sylv'ACCTES, 2018b). Consequently, Sylv'ACCTES benefits from a system of tax exemption delivered by the fiscal administration. Hence, the organization is allowed to deliver tax receipts that provide a 60% tax exemption for any donation. In other words, a company giving 100,000 € only incurs a 40,000 € expenditure after the tax credit. Application for membership is once again subject to an ethical committee.

Whether NF and Sylv'ACCTES like it or not, their funders are also attracted by the opportunity to improve other forest ES through the carbon offset contracts. Business funders viewed Sylv'ACCTES and NF as an opportunity to offset their carbon emissions and simultaneously strengthen their social and environmental responsibilities at a local level. The corporate funders of Sylv'ACCTES are mainly motivated by the positive social and economic consequences of forestry operations, emphasized as one of the reasons for participating in the regional development of the forestry sector: "Bringing together private firms in or within territories: this is a stronger message than saying 'you love forests, so give to forests'" (S11) or "What we want to do is to create a short circuit. We have the local authorities, local foresters, and local companies accompanying the management of a natural resource toward something more virtuous" (S1).

For regional and local authorities, NF and Sylv'ACCTES open an additional avenue to reinforce their forest-based policies such as the economic development of the timber industry, the reinforcement of the renewable energy sector, the promotion of open-air leisure activities, and so on. This holds particularly true regarding the increased wood mobilization, which is an important motivation for regional councils and cities to fund carbon organizations. The regional branch of the Agency for the Environment and Management of Energy (ADEME), which was instrumental in the launch of NF, explained: "In the end, mitigation is not everything. For us, it's just a good reason to influence other fuelwood policies" (NF3). At a smaller scale, several cities in Normandy joined NF, as they are currently developing collective heating facilities: "We wanted to mobilize a lot of extra wood, mobilize wood-fired heating plants in private forests" (NF3). For Sylv'ACCTES, the regional need for fuelwood heating was one of the reasons for public funders to become involved. The Auvergne–Rhone–Alpes region has a strong culture of fuelwood heating that continues to develop, and it encouraged an organization that secured the local and sustainable production of fuelwood: "The 'Grand Lyon' metropolis had problems with provisioning their wood-fired heating plants. We also worked with territories on timber production, so there was this interesting rural-urban link" (S10). Along with fire heating, the rural–urban link was depicted by both associations as a communication support that strengthened the relationships between forest operators, local companies, and inhabitants: "From the perspective of Sylv'ACCTES' governance, its major interest lies in the creation of a local dialogue" (S7).

3.3. Sustainable Timber Production: A Key Motivation Underlying Offset Contracts

Along with carbon storage and other forest ES, increasing or securing timber production emerged as a key underlying motivation in the creation of new forest intermediaries oriented toward carbon storage. The involvement of the Regional Centers for Privately Owned Forests (CRPF in French) in Normandy and Auvergne–Rhône–Alpes was a determining factor in the foundation of NF and Sylv'ACCTES, as mentioned on their websites, in many interviews, and in the NF offset contracts; indeed, CRPFs are public bodies that promote the indirect stimulation of the timber industry through increased wood mobilization in private forests [31]. For NF especially, producing wood was a key issue, since regionalized forest growth models can predict a gap in wood production due to the unbalanced age structure of forest stands. To supply the missing age category, the Normandy CRPF viewed NF as a relevant interface to pay for the clear-cut and replanting costs of decaying and unproductive stands, thus avoiding a potential problem linked to timber and fuelwood provisioning in the coming decades.

In line with the CRPFs, private landholders and public forest managers who enter into an offset contract consider the funding of Sylv'ACCTES and NF to be a relevant solution for funding timber production. This holds true for the funding of forest operations that would otherwise not occur: "For us, the starting point was mobilizing the wood of poor forest stands" (NF2). Yet it also applies to operations aimed at improving the quality of harvested wood: "The idea of Sylv'ACCTES is to really improve the value of the timber" (S5). In both cases, the funding amount of the contracts is not intended to fully cover the costs of the forest operations but is rather seen as an incentive for forest landowners to set them up, as underlined by interviewee NF3: "The amount was evaluated so that it would be large enough to commit to the program." To determine how much of the costs of forest management would be reimbursed, Sylv'ACCTES decided to focus on forest expenditure independently of carbon markets, with the offset contracts funding 40% of the operations. By contrast, NF decided to set the amount of funding based on the market price of a carbon ton and calculated that each hectare replanted after clear-cutting could store a certain amount of carbon, equivalent to 2000 € per hectare (capped at a maximum of 30% of the costs). NF explained that the funding requirements should be more attractive than public subsidies: "It has to be simple and not supported by EAFRD [European Agricultural Fund for Rural Development], relatively fast in reaching a decision, and exploitable for the landowner" (NF2).

While timber production was an important issue in the relations between forest landholders and the carbon organizations, Sylv'ACCTES and NF remained vigilant about its sustainability, especially regarding the protection of natural habitats. To be eligible for the offset contracts, forest landowners had to comply with environmental prerequisites. As proof of commitment to biodiversity, NF and Sylv'ACCTES require an official document that is mandatory in France for properties over 25 ha (and optional below), certifying that the target stands are sustainably managed. As an additional environmental condition, NF requires a sustainable management certification delivered by the Program for the Endorsement of Forest Certification (PEFC) or the Forest Stewardship Council (FSC). Sylv'ACCTES does not demand environmental certifications for a forest stand but instead requires that at least 30% of its surface has a species mixture, with at least five old-growth trees per hectare. To monitor the ecological impact of its offset contracts, Sylv'ACCTES patented an indicator known as "Potential Biodiversity Gain" (www.sylvacctes.org/les-indicateurs-sylv-acctes/, (accessed on 5 October 2018). The indicator developed from early discussions between Sylv'ACCTES and members of environmental organizations, who were invited to be part of the organization's scientific committee and contribute to the design of the environmental specifications of the offset contracts.

To improve the funding success of forest operations, Sylv'ACCTES and NF tried to reduce the detrimental effects of timber production on the contribution of the forests to the landscape. Several interviewees mentioned their fear of the social pressures linked to forestry operations, which is commonly expressed by inhabitants in their rejection of tree harvesting. Interviewee NF2 described a public conference on forestry held in Normandy: "Several people expressed their feelings for the trees, and we saw that many people were there to defend the trees and stop an overly interventionist approach by limiting the number cut down." In anticipation of such claims, the communication strategy of NF stressed the positive outcomes of forestry operations for carbon mitigation or local employment, and not only the technical aspects. Information boards were displayed at the site of each supported operation, and forest owners who signed the offset contracts also acknowledged the right of funders to visit the forest stand.

4. Discussion

NF and Sylv'ACCTES have emerged as new intermediary forest organizations, aiming to bring together forest managers (carbon storage providers) and forest ES beneficiaries around carbon storage projects. Although the formal institutionalization of NF and Sylv'ACCTES is undeniable, their strategy to manage efficient, sustainable, and multifunctional projects for forest carbon storage must be questioned in the light of the actual outcomes. Figure 2 illustrates the discussion using the conceptual framework of Barnaud et al. for ES, social interdependencies, and collective action.

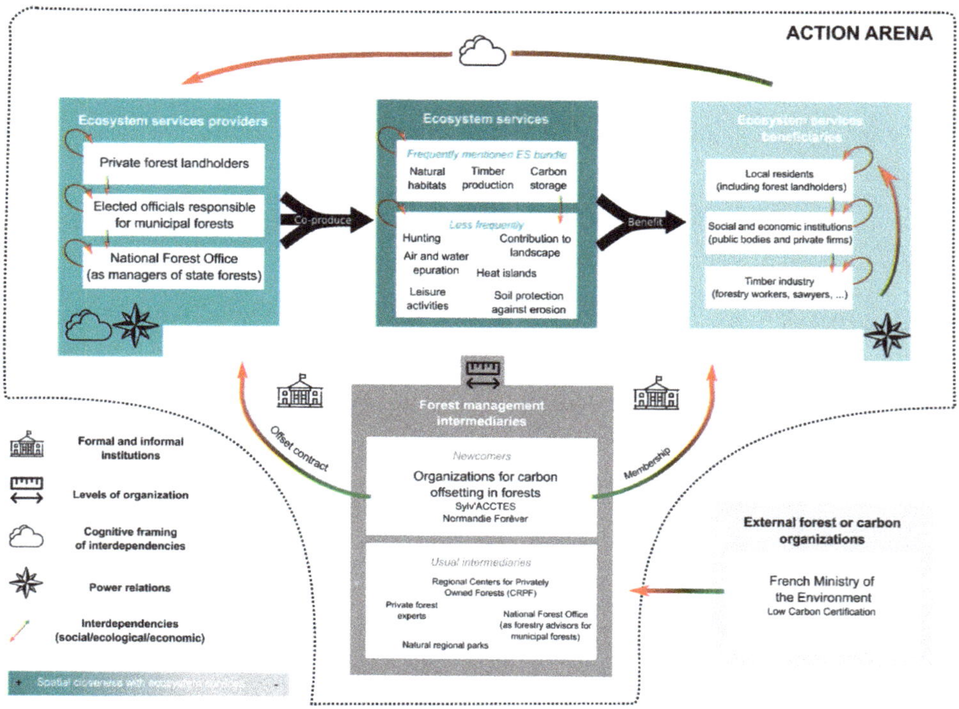

Figure 2. The conceptual framework for ecosystem services (ES), social interdependencies, and collective actions applied to French forest socio-ecosystems.

4.1. On the Efficiency of Carbon Storage Organizations

We posit that the efficiency of the carbon projects initiated by NF and Sylv'ACCTES does not lie in their new methodological approach to overcome the four challenges of carbon storage but rather in their design of operational contracts to increase the number of participating landholders. Conditioning the financial aid to an obligation of means (technical specifications) and not to an obligation of results (reaching a given amount of stored carbon) was a strategic choice of both NF and Sylv'ACCTES to avoid deterring forest ES providers, which enabled them to efficiently recruit forest managers.

Even so, NF and Sylv'ACCTES chose three proven solutions to design forest contracts that efficiently overcome the usual challenges of carbon storage. Based on the recommendations of their scientific committees, they opted for a "physical" buffer system as well as ex-ante and in-itinere verifications. The physical buffer system reports less carbon storage than initially calculated, which is a common means to prevent heterogeneity and uncertainty issues. A carbon reserve buffers deviations caused by heterogeneous and uncertain forest conditions based on the difference between the amount of carbon stored after the contract closure and what was anticipated in the reference scenario [7,23,34]. Uncertainties in the calculation parameters can stem from their inherent stochasticity (e.g., local climate), entanglement (e.g., tail dependence effects strengthened by climate change; [15]), a lack of reliable data (e.g., soil condition, browsing pressure), not to mention the knowledge gap regarding phenomena that are only now coming to light (e.g., storage capacity of upper soil horizons or old-growth versus recent forests; [35–37]). The second solution prevented additionality issues. Both organizations chose an ex-ante evaluation of applicants' motivations to sign the offset contracts. This case-by-case assessment aimed to discard candidates who would conduct forest operations regardless of the offset contracts. The two organizations also planned in-itinere verifications of the permanence of the subsidized projects. Their approach, used in many other countries [23,38,39], aims to include the transaction costs of impermanence verifications in their estimated budgets. The organization either funds an external certifying body such as the Verified Carbon Standard [40] or pays for the costs of verifying the contracts itself. Only NF introduced a replanting clause in its contract in the event of destruction following a natural hazard, which was certainly motivated by its objective to prepare the timber industry for the upcoming decades. The lack of ex-post verifications of the efficiency of the offset contracts (for instance, by measuring how much additional carbon will be stored) underlines the preference of NF and Sylv'ACCTES for means-oriented, not results-oriented, offset contracts.

The preference of NF and Sylv'ACCTES for means-oriented contracts emphasizes their strategy of focusing on the second limiting factor of offset contracts, namely the reticence of landholders to commit to their program. Subsidies for forestry operations can be scarce and sometimes so complex that they deter many forest owners from applying for them. Both public foresters and private owners struggle to balance their management expenses with timber incomes. Public foresters working for the National Forests Office are in charge of the management of national forests but have to reduce their expenditure [41]. Private foresters do not strive for profit at any cost, although they can no longer rely on traditional subsidies to keep their budget balanced [42]. Known as the National Forest Fund, a funding program was launched in the aftermath of World War II to boost wood production in order to rebuild the country. However, this program ended in 2000, thus inducing a loss of financial incentives for private owners [43]. The solution of the offset contracts proposed by NF and Sylv'ACCTES partially overcomes the economic impediment to forest management. Compared with the national tax exemption known as DEFI (French acronym for "Dispositif d'Encouragement Fiscal à l'Investissement", meaning "fiscal measures supporting investment") piloted by the Ministry of Agriculture, which is also in charge of forests, Sylv'ACCTES contributes twice as much to the costs of sylvicultural operations (until December 2020, DEFI could refund up to 6250 € per capita for forest operations expenses [44]). Its engagement period is five years shorter (10 years vs 15 years for DEFI), and its minimum surface is half the size (2 ha vs. 4 ha for DEFI) [33]. Another key feature of the offset contracts is that they abide by the financial concerns of landowners and base the amount of the subsidy on the cost of the forestry operations. Strictly correlating an aid to international market prices for a ton of carbon dioxide would have created uncertainty and discouraged landowners, as observed in many REDD+ projects [40]. The means-based design of the offset contracts of NF and Sylv'ACCTES ensured their legibility and visibility, thus increasing their appeal to forest owners. However, in spite of these attractive conditions, it appears that subsidy contracts are not given full consideration by private owners, who are a major target of NF and Sylv'ACCTES. Indeed, private forests cover 75% of forested areas in France, with the remainder being split between state forests and public entity forests [31]. One explanation may relate to the reluctance of private owners to engage in long-term contracts, even for 10 years, as observed in other voluntary forestry programs such as conservation programs [45]. Another non-exclusive explanation is that the French forest owners most interested in carbon mitigation tend to be the youngest and most educated ones [21], an underrepresented group among private forest owners [46].

The efficiency of the offset contracts in recruiting forest ES providers must therefore be assessed in the light of the recipient owners. Public forest managers are overrepresented, which reveals another obstacle to the activities of NF and Sylv'ACCTES: the long-standing problem of the ownership structures of private forests. The fragmentation of private forests, the ageing of their plentiful owners, and the share of absentee forest owners are internationally known to impede private investment in carbon offset contracts [38,39], and this equally applies to the French forest sector [46]. The efficiency of NF and Sylv'ACCTES to connect with forest ES providers is therefore not yet fully exploited, as both organizations mainly attract public forest managers. Municipalities owning forests find the offset contracts to be efficient. As they often struggle to balance their forest budgets, these municipalities welcome this additional funding source. Besides, elected officials now have an alternative forestry advisor with carbon organizations compared with their usual interaction with the National Forest Office. The new possibility of sharing critical financial, technical, or administrative knowledge about forest management reinforces elected officials' power of decision. In this regard, our findings further evidence how the emerging intermediaries of forest carbon storage impact the power distribution among forest ES providers.

At the time of writing, there were no ex-post evaluations of the offset contracts of NF and Sylv'ACCTES. Due care must therefore be paid to assessing their overall effectiveness in enhancing forest carbon storage. Yet we strongly believe that their science-based technical approach is certainly not counterproductive to enhancing carbon storage. On the contrary, it is presumably the best approach to efficiently recruit forest ES providers in times of uncertainty, especially since the means-oriented (and not results-oriented) conception of the offset contracts does not fundamentally change the distribution of economic power among FSES stakeholders, thus avoiding potential conflicts.

4.2. *Sustainability of Carbon Offsetting Dependent on Its Multifunctionality*

Multi-objective forest operations are most often inconsistent with a management approach geared toward only one ES such as carbon storage, unless an appropriate spatial resolution is found [47]. This situation is commonly illustrated with management intensification for timber production, which can contradict biodiversity conservation or water quality preservation, for instance [8,27–29]. Consequently, NF and Sylv'ACCTES paid attention to the environmental, social, and economic outcomes of forest management to avoid conflicts of use with other forest ES providers or beneficiaries.

To ensure their environmental sustainability, NF and Sylv'ACCTES called on environmental partners to help develop clauses for environmental safeguards in the offset contracts. NF required a sustainable management certification delivered by PEFC or FSC, a widespread practice used to account for environmental concerns [48]. Sylv'ACCTES does not use environmental certifications but instead requires a species mixture covering at least 30% of the forest surface and at least five old-growth trees per hectare. Neither of these criteria proved to be challenging for forest owners, as many stands already meet these requirements in the Auvergne–Rhône–Alpes region [31]. We assumed that establishing such environmental safeguards was a genuine requirement for NF and Sylv'ACCTES and a key to strengthening the communication strategy around carbon mitigation and avoiding any pressure from environmental organizations.

The social sustainability of NF and Sylv'ACCTES relies on the non-partisan cognitive framing of the offset contracts, portrayed as a winning combination of various interests that goes beyond party lines. This ensured the stable launch and expansion of the organizations, as illustrated at the very beginning of Sylv'ACCTES. A political shift at the head of its region could have halted the financial contribution of the regional council, but the newly elected regional officials maintained their support. In our opinion, the non-partisan cognitive framing of the offset contracts went hand in hand with the communication efforts emphasizing the neutral or synergetic rearrangement of social relationships due to the forest operations. Involved in only a limited manner with the nonprofit organizations, hikers and local inhabitants could read information boards championing the positive role of forest management in providing them with forest ES (especially carbon storage). This defused any potential conflict with these beneficiaries of the landscape and recreational forest ES. When more actively involved, the forest owners increased their awareness of their contribution to the mitigation of climate change. This socially rewarding position was prone to avoid conflict and even convert them into supporters. Similarly, the attention paid to the environmental sustainability of the contracts and the communication centered around their positive economic and regional outcomes were key aspects for securing the recruitment of new funders. This started a virtuous circle in which the positive labeling of local businesses and public entities (e.g., mayors) that supported NF and Sylv'ACCTES boosted the arrival of new funders from beyond the "usual" forest ES providers. This strategy of NF and Sylv'ACCTES—"prevention (of conflict) is better than cure"—is also illustrated by their way of dealing with the greatest risk of social conflict, namely the current configuration of the offset contracts that rewards the "black sheep." To avoid the usual funding of forest stands with the worst practices of GHG mitigation, eligibility criteria might progressively change to follow the evolution of mitigation policies—with a risk of subsequently lessening the interest of forest owners in this type of incentive.

The positive outcomes of the offset contracts on social and economic forest ES were strong incentives to attract public bodies and private companies eager to voluntarily offset their GHG emissions. They also contributed to securing support from the traditional forest management operators. The respect for the funders' core values and organizational self-interest is known to expand membership [49], a mechanism even bolstered by the inclusion of environmental safeguards to reduce the reputational risks of "greenwashing." By broadening the range of potential funders to private businesses and public entities, the carbon organizations need to consider the multitude of forest ES to unlock additional financial sources based on mitigation. In the case of Sylv'ACCTES, this can be seen through the recognition of general interest, which does not acknowledge forest management but is rather motivated by the "defense of the natural environment." The tax exemption was a strong lever to ensure the funding of the organization's subsidy system. To the best of our knowledge, this type of financial structure is unknown in the forestry sector and may be expanded, as stressed by the desire of NF to be recognized to be of public interest. However, we argue that the support of both public and private funders is a powerful lever for funding forest management, although it is still double-edged. Concerns can be raised regarding the gradual development of asymmetric power interdependencies between ES providers and ES beneficiaries, with a shift of forestry funding from the EU and state-based subsidies to private and regional-based funding. Voluntary contributions to the funding of forest management should not exceed a certain threshold in the eventuality of a funder withdrawing support, because it could or would not maintain funding. The same applies to NF, for which offset contracts should first aim at solving the regional timber issue; hence, development further afield is currently not given much attention, or at most, in neighboring regions. Funding issues could therefore arise once this problem is solved.

4.3. Territorial Scale Conditioning the Efficiency and Sustainability of Carbon Offsetting

As raised in most interviews, the territorial scale of the carbon organizations was deemed essential for their efficiency and sustainability. Both NF and Sylv'ACCTES achieved a specific cognitive framing around the social-ecological interdependencies between forest ES providers and carbon offsetters. For instance, they outlined the regional benefits of forest climate services that connected local funders and forest ES providers (e.g., [50]). In other words, they reduced the social distance between them, which is an important feature of bottom-up collective action by decreasing the transaction costs of coordination [51]. For instance, the original clustering of the two carbon organizations with public entities and CRPFs brought a form of insurance against potential greenwashing allegations against the private funders. A consensus is now emerging in the carbon organizations regarding the importance of regional embedding for their development, which explains the caution shown in terms of expanding to other regions so as not to lose operationality. Indeed, while collective action operates well when social distances are small, scaling-up to higher spatial levels of organization favors market- and state-based institutions [30]. The same applies to the local adaptation of the concept of multifunctional forestry [47,52].

The issue of securing current projects and expanding is now arising, especially since a national LCC was portrayed as an incentive for voluntary mitigation projects. Inspired by other EU initiatives such as the woodland carbon plan in the UK [53], the Ministry of the Environment tasked the semi-public Institute for Carbon Economics (I4CE) with developing a voluntary mitigation certification known as the "low carbon certification," which was created to contribute to the Ministry's carbon neutrality policy for 2050. Officially released in November 2018 by the Ministry of the Environment, the LCC consists of a checklist of general features similar to international standards [13]. According to the LCC, offsets must be real, measurable, verifiable, and supplementary [54]. The LCC is also supposed to reject projects with "potentially significant negative impacts from an environmental and socioeconomic perspective" [51]. Broadly, the LCC encourages bottom-up initiatives. For the forestry sector, any field practitioner can launch a project. Yet three major limitations in the LCC process arise from the rifts between top-down international mitigation standards and field-based indicators. First, as with Sylv'ACCTES and NF, the results-based additionality, exactitude, and exhaustivity principles are too demanding given the numerous knowledge gaps and uncertainties surrounding carbon mitigation. Second, the engagement period of 30 years was chosen in line with the 2050 target for carbon neutrality, but this does not align with the temporality of foresters nor with the need for businesses to assess the impact of their funding in timeframes that are compatible with their CSR policies. This is a problem for organizations such as NF, which aim for a positive carbon balance in 70 years instead of 30 years. Third, as explained in its official description, no financial incentives are guaranteed by the certification [51]. The only incentive for using the LCC is gaining a good reputation once granted. Paradoxically, the certification encourages bottom-up approaches that nurture mutual trust between project participants, the same trust that absolves them from relying on external certification. Similar observations have been made in other national contexts [40].

Only the coming years will tell if locally based mitigation projects will require the LCC to provide a guarantee to potential corporate funders with little knowledge of offset processes. This would lead to a switch from local funders located near the carbon organizations to nationwide firms trying to offset their GHG emissions using the LCC to anticipate and avoid future regulatory burdens or attract investors [49]. We argue for the coexistence of bottom-up carbon organizations and carbon certifications such as the LCC, so as to complement (and not replace) means-oriented forest intermediaries with results-oriented mitigation tools. The EU, for instance, places a strong emphasis on reporting the results obtained from subsidies, with less importance given to the means used [55].

5. Conclusions

The outcomes of global changes on forest management are manifold, as illustrated by our case study on the plural effects of GHG mitigation. We outline here how offset contracts succeeded in cementing bonds between professionals tasked with the development of forestry operations, private and public structures interested in GHG mitigation, and landowners struggling to fund the management of their forests. In our opinion, the two carbon organizations did not innovate in terms of the technical operations adopted to store more carbon in forests—they simply retained the best practices known to date. Their most striking feature is the development of collective action in the FSES based on a forest ES other than timber production.

A few key points can be learned from this study and used for the development or evaluation of further collective actions of carbon storage in FSES.

First, we argue that, regardless of the national or regional context, considering the singularities of the FSES—especially its multifunctionality—and its stakeholders is a key element. Indeed, the best way to strengthen voluntary bottom-up projects is to build trust by protecting the interests of members and their work culture [51]. The efficiency and sustainability of the intermediary forest organizations therefore depend on integrating the carbon mitigation challenges with the issues of concern of their contributors and beneficiaries. Here, we recommend carbon organizations to specify the main objectives of their carbon offsetters and beneficiaries: are they willing to strengthen their environmental responsibility? Do they want to open a new funding channel for forestry operations? How long are they prepared to participate? Etc.

Second, the social and economic sustainability of the offset contracts also relies on the prevention of conflicts among and between forest ES providers, forest ES beneficiaries, and other forest management intermediaries (e.g., natural regional parks). Conflict prevention also stemmed from social learning about the interdependencies of all the actors in the action arena [56]. While the FSES can be susceptible to various open conflicts because of the incoming subsidies [40], our case studies encountered relatively few confrontations despite tensions around the prioritization of forest ES. To avoid an explosive situation in which timber production or environmental conservation is hindered by the offset contracts, NF and Sylv'ACCTES worked for collective governance to balance the power distribution between FSES stakeholders. By bringing together the representatives of the different groups of the FSES (e.g., members of environmental organizations, elected officials), they opted for what may seem to be a "weak consensus" regarding multifunctionality. According to this vision, the enhancement of forest ES (carbon storage) stops where another (timber production, habitat preservation) begins. More specifically, we recommend proactively inviting any stakeholder showing interest in the carbon offsetting to take part in the process [5]. If required, distinct discussion arenas can be initiated to prevent the re-emergence of long-standing conflicts (e.g., ethical, technical, scientific committees).

Third, the intermediary carbon organizations made significant efforts for the cognitive framing of the offset contracts, thus proving their added value for every protagonist of the forest action arena. We assume that institutionalizing the link between carbon offsetters and public and private forest landholders is one of their main achievements. This exemplifies the social–ecological interactions between forest stakeholders across spatial and temporal levels. We posit that maintaining effective internal and external communications about the objectives, processes and results of forestry operations is paramount in boosting forest management for carbon offsetting.

We also draw the reader's attention to the potential weaknesses of the process. The strategy of enhancing carbon storage by effectuating small changes to forest operations on a maximum of different forest plots only works if multifarious forest owners and managers can be targeted. This strategy requires both a good knowledge of the structure of forest ownership [38] and communication skills to avoid presenting offset contracts as a panacea to fund forest operations. Indeed, foresters have shown their interest in the financial initiatives for environment services, especially in a context of rising climatic hazards. Currently, French forest landholders generate most of their forest-based revenues from timber sales (and in some cases, the lease of hunting rights), but these are at risk with the increase in climate hazards. It could be important to keep in mind that carbon subsidies are not intended to cover all the expenditure of forest management but rather to act as a lever for forestry operations along with incomes stemming from the valuation of other forest ES. Lastly, the state government should encourage voluntary initiatives to offset GHG emissions, although they should not forget that they are means-oriented, which contradicts their broader integration into results-oriented mitigation policies (e.g., REDD+ [16]).

Time is now needed to assess the actual contribution of these voluntary actions to the overwhelming issue of global climate change. However, they seem to be robust and serious, as shown by the consideration granted to forest operations that acknowledge the adverse effects of climate change. As a sign of the times, the effectiveness of intermediary organizations that promote carbon storage has gained national and international recognition, as shown by the presentation of Sylv'ACCTES as a "nature-based solution" by the International Union for Conservation of Nature [57].

Author Contributions: Conceptualization, T.F. and L.G.; methodology, T.F. and L.G.; validation, M.T., N.F.-L.; formal analysis, T.F. and L.G.; investigation, T.F. and L.G.; resources, T.F. and L.G.; data curation, T.F. and L.G.; writing—original draft preparation, T.F. and L.G.; writing—review and editing, M.T. and N.F.-L.; visualization, T.F.; supervision, N.F.-L.; project administration, T.F.; funding acquisition, N.F.-L. All authors have read and agreed to the published version of the manuscript.

Funding: This research was funded by the French National Research Agency (ANR), grant number ANR-11-LABX-0034 as part of the "Investissements d'Avenir" Program (LabEx BASC).

Institutional Review Board Statement: Not applicable.

Informed Consent Statement: Informed consent was obtained from all subjects involved in the study.

Data Availability Statement: The data presented in this study are available on request from the corresponding author. The data are not publicly available due to privacy reasons.

Acknowledgments: The authors are grateful to the members of Sylv'ACCTES and Normandie Forêver for their willingness to share their knowledge. The article was improved thanks to the comments of two anonymous reviewers, to whom we are very grateful.

Conflicts of Interest: The authors declare no conflicts of interest. The funders had no role in the design of the study; in the collection, analyses, or interpretation of data; in the writing of the manuscript, or in the decision to publish the results.

Appendix A

Table A1. List of identified carbon projects. All websites were accessed online on September and October 2018.

Organization Name	Source
Coeur de Forêt	https://www.coeurdeforet.com
Collectif Bois 07	https://collectifbois07.wordpress.com
Duramen	https://www.duramen.org
EcoTree	https://ecotree.green
ERE43	https://www.ere43.fr
Normandie Forêver	http://www.normandieforever.org https://www.facebook.com/normandieforever
Pur Projet	https://www.purprojet.com
Reforest'Action	https://www.reforestaction.com
Sylv'ACCTES	https://sylvacctes.org

Appendix B

The overall objective of the interview grid was to describe the "forest carbon storage" action arena using the following categories:

- Structure and functioning of the organization: Does the carbon offsetting involve technical changes (e.g., new tools, sylvicultural management practices) or logistic and organizational changes (e.g., new organization in the relations and administration of forest industry)?
- Efficiency of carbon storage: What are the results of the projects? What are the indicators and references used to measure carbon storage (e.g., consideration of both above- and belowground carbon compartments)?

- Sustainability of the organization: Is biodiversity an issue? If yes, how is it included in the forest operations geared toward carbon storage? How are the financial subsidies determined? Are they calculated based on carbon markets?
- Consideration for forest multi-functionality/cognitive framing: Along with carbon storage, which other ecosystem services/forest functions/uses are targeted? Which results matter the most? Do the different forest stakeholders and carbon offsetters share the same vision of forest management?
- Spatial and temporal dimensions of the organization: Will the organization endure in time (e.g., climate uncertainties) and space (i.e., can it be replicated in other territories, and if yes, what are the specific methodological tools for a given territory)?
- Stakeholders' interactions within the organization: How are the relations between actors organized? Is there anything novel about the way in which carbon offsetters and foresters interact (compared to the usual action arena of French forestry)? If yes, what is it?
- Power relations: How are the carbon storage organizations structured? Where does the executive power lie? Do they encounter the same deadlocks as those emphasized by traditional timber growers?

Table A2. Overview of the interviewees.

Project	Anonymization Number	Sex (Male/Female)	Role in the Offset Project
Sylv'ACCTES	S1	M	Cofounder and coordinator
	S2	F	Management of a forest where an offset project takes place
	S3	F	Member of the technical and scientific committee
	S4	M	Management of a forest where an offset project takes place
	S5	F	Management of a forest where an offset project takes place
	S6	F	Management of a forest where an offset project takes place
	S7	M	Member of the technical and scientific committee
	S8	M	Member of the technical and scientific committee
	S9	M	Funder (carbon offsetter)
	S10	F	Cofounder
	S11	M	Funder (carbon offsetter)
Normandie Forêver	NF 1	M	Cofounder and coordinator
	NF2	F	Comanager of a forest where an offset project takes place
	NF3	M	Cofounder and funder

Table A3. Grid of the semi-directed interviews.

Interview Category	Questions	Alternative Formulations to Relay the Question
Interviewee's relation to the carbon offsetting organization	- Which professional or life path brought you to work with/in [name of the organization]? - How did you become involved in [name of the organization]? - What actions have you initiated since joining the project?	- Could you talk about the beginning of the project and its implementation? Or about the moment when you arrived on the project? - What attracted you to this project? - Which projects did you conduct? Which decisions did you or do you regularly take? - What have you done with [name of the organization] that you were unable to do elsewhere?
Interviewee's relation to forest management("cognitive framing")	- Do you see any links between forests and climate change? - What are the common objectives of forest management for you? - (For forest experts and owners) - Which professional or life path brought you to work with/in the forest sector? - How do you usually work in the forests that you manage/own?	- What are the forest operations that you usually do? - Which outcome(s) have you chosen for your timber? - (forest experts) - How do you deal with the demands of the general public/forestry sector? - Which type of silviculture do you practice? (e.g., mixed/monospecific stands, (ir)regular shelterwood) - (Forest owners) - How did you choose your forest? - Who did you choose to manage your forest?
Project and its development	- How did the idea of [name of the organization] emerge? - Did a founding moment occur? - What were the building steps?	- Have you noticed anything in the timber industry that could be improved or that drove you toward this project? - What was the identified need that gave rise to the project? - What was the initial goal? Is it still the same? - (For [name of the organization] staff members) - Who are the people working with you? - Based on which criteria did you choose them? - How were the forestry experts chosen in the organization? How are the relationships between the different experts?

Table A3. *Cont.*

Interview Category	Questions	Alternative Formulations to Relay the Question
ResultsFunctioning	- How do you monitor the results of the project? - Have you fulfilled the initial objectives made when you joined the project? How can you measure them? - Have you noticed any evolution regarding the initial objectives or since you arrived? - What does the regular progress of a project look like?	- What are the results? Are you satisfied? - How did you respond to the identified need during the project design? - Can you talk about your impact assessment (tools, numbers, studies)? - What is your strategy for contacting the different actors of a new project? - How did you raise funding? - Could you describe a typical month for [name of the organization]? - (Territorial anchorage): At which scale did you initially design the project, and why?
Sustainability	- How do you monitor the reduction of your carbon impact? - How do you assess the biodiversity dynamics (before and after the implementation of the project)?	- If your forests are already registered in a sustainable management plan, what could be optimized in terms of their carbon impact and biodiversity? - Which method do you use to calculate your carbon impact in forestry, and which factors do you include in your method? - Please describe your strategy to reduce your carbon impact. How does it work? Does it focus on carbon sequestration, storage, or substitution? - On which environmental aspects do you focus the most, and how do you chose your priorities? (Mention biodiversity at this point if the interviewee has not yet done so).
Project's capacity to help actors interact with one another	- With whom do you work, and what does it practically mean? To whom do you talk about the carbon project? - How often do you meet your counterparts?	- (Personalize the questions depending on the interviewee's connection with the carbon project): - In your view, why is it important to work with the forestry sector/corporations/public institutions? - Do you think that [name of the organization] connects different actors that rarely work together or interact? - Do you think that you can help the timber industry in a new way (e.g., hiring forest workers)?
Potential evolution of the project	- Do you have plans for future developments? - Do you know about any similar initiatives with the same objectives? - How are you evolving in the national context of multiple climate policies?	- How do you see the project evolving in the coming years? - What are your objectives for the years to come? - Do you consider expanding to new territories? If yes, which ones and why? - Do you think that your project is more suitable for a particular type of territory, or can it be used everywhere? Can it be applied to ecosystems other than forests? - Did you contribute to the national adaptation plan, mitigation strategy, etc.?

Appendix C

Analysis grid for the transcribed interviews.

Appendix C.1 Objectives/Issues of the Organizations ("Why"?)

Appendix C.1.1 Issues Specific to Only One Organization/Code #1

- Normandie Forêver (NF): forest dieback (for foresters and beneficiaries)
- Sylv'ACCTES: sustainable management of forests, wood production, relationship with the public

Appendix C.1.2 Issues Common to Both Organizations/Code #2

- Content of the offset contracts (biodiversity or social development clauses)
- Managing private forests areas (gathering together private owners, leading them toward a more efficient and sustainable forest management)
- Need for fuelwood energy supply

Appendix C.2 Organizational Processes ("How"?)

Appendix C.2.1 Organizational Processes Specific to Only One Organization/Code #3

- Funding methods

Appendix C.2.2 Organizational Processes Common to Both Organizations/Code #4

- Distribution of decisional power between stakeholders

- Actors involved (private-public partnership)
- Jurisdictional form as nonprofit organizations
- Spatial and temporal scale > local development

Appendix C.2.3 Results Obtained/Code #5

- Quantitative results > area covered, money collected, future perspectives

Appendix C.3 Carbon Mitigation in the Organization's Development

Appendix C.3.1 Cognitive Framing of the Process/Code #6

- NF > main goal is carbon mitigation, with firms paying for the restoration of the forest dieback and thus contributing to carbon mitigation
- SACC > carbon mitigation is one of the positive externalities of sustainable forest management, BUT it is an appealing argument to carbon offsetters

Appendix C.3.2 Calculation of Carbon Storage/Code #7

- Methodology used
- Control of results obtained

References

1. Seidl, R.; Thom, D.; Kautz, M.; Martin-Benito, D.; Peltoniemi, M.; Vacchiano, G.; Wild, J.; Ascoli, D.; Petr, M.; Honkaniemi, J.; et al. Forest Disturbances under Climate Change. *Nat. Clim. Chang.* **2017**, *7*, 395–402. [CrossRef]
2. Keenan, R.J. Climate Change Impacts and Adaptation in Forest Management: A Review. *Ann. For. Sci.* **2015**, *72*, 145–167. [CrossRef]
3. Huang, L.; Zhou, M.; Lv, J.; Chen, K. Trends in global research in forest carbon sequestration: A bibliometric analysis. *J. Clean. Prod.* **2020**, *252*, 119908. [CrossRef]
4. Mayer, M.; Prescott, C.E.; Abaker, W.E.A.; Augusto, L.; Cécillon, L.; Ferreira, G.W.D.; James, J.; Jandl, R.; Katzensteiner, K.; Laclau, J.-P.; et al. Tamm review: Influence of forest management activities on soil organic carbon stocks: A knowledge synthesis. *For. Ecol. Manag.* **2020**, *466*, 118127. [CrossRef]
5. Fouqueray, T.; Frascaria-Lacoste, N. Social sciences have so much more to bring to climate studies in forest research: A French case study. *Ann. For. Sci.* **2020**, *77*, 81. [CrossRef]
6. Corbera, E. Problematizing REDD+ as an experiment in payments for ecosystem services. *Curr. Opin. Environ. Sustain.* **2012**, *8*, 612–619. [CrossRef]
7. Blanc, S.; Accastello, C.; Bianchi, E.; Lingua, F.; Vacchiano, G.; Mosso, A.; Brun, F. An integrated approach to assess carbon credit from improved forest management. *J. Sustain. For.* **2019**, *38*, 31–45. [CrossRef]
8. Duncker, P.S.; Raulund-Rasmussen, K.; Gundersen, P.; Katzensteiner, K.; De Jong, J.; Ravn, H.P.; Smith, M.; Eckmüllner, O.; Spiecker, H. How forest management affects ecosystem services, including timber production and economic return: Synergies and trade-offs. *Ecol. Soc.* **2012**, *17*. [CrossRef]
9. Fouqueray, T.; Charpentier, A.; Trommetter, M.; Frascaria-Lacoste, N. The calm before the storm: How climate change drives forestry evolutions. *For. Ecol. Manag.* **2020**, *460*, 117880. [CrossRef]
10. Borrass, L.; Kleinschmit, D.; Winkel, G. The "German model" of integrative multifunctional forest management—Analysing the emergence and political evolution of a forest management concept. *For. Policy Econ.* **2017**, *77*, 16–23. [CrossRef]
11. Cubbage, F.; Harou, P.; Sills, E. Policy instruments to enhance multi-functional forest management. *For. Policy Econ.* **2007**, *9*, 833–851. [CrossRef]
12. Hommel, T.; Godard, O. Que peut-on espérer des entreprises socialement responsables? In *Regards Sur la Terre 2009*; Annuels; Presses de Sciences Po: Paris, France, 2009; pp. 167–178. ISBN 978-2-7246-1091-8.
13. Hamrick, K.; Gallant, M. *Voluntary Carbon Markets Insights: 2018 Outlook and First-Quarter Trends*; Ecosystem Marketplace, Forest Trends: Washington, DC, USA, 2018.
14. Bertolas, R.J. Cross-Cultural environmental perception of wilderness. *Prof. Geogr.* **1998**, *50*, 98–111. [CrossRef]
15. Cooley, D.M.; Galik, C.S.; Holmes, T.P.; Kousky, C.; Cooke, R.M. Managing dependencies in forest offset projects: Toward a more complete evaluation of reversal risk. *Mitig. Adapt. Strateg. Glob. Chang.* **2012**, *17*, 17–24. [CrossRef]
16. Phelps, J.; Webb, E.L.; Agrawal, A. Does REDD+ threaten to recentralize forest governance? *Science* **2010**, *328*, 312–313. [CrossRef] [PubMed]
17. MAAF. *Programme National de La Forêt et Du Bois 2016–2026*; Ministère de l'Agriculture, de l'Agroalimentaire et de la Forêt: Paris, France, 2017; p. 60.
18. Barthod, C. La Multifonctionnalité des forêts entre discours et pratiques: Illusion ou réalité à assumer? *Rev. For. Française* **2015**, *4*, 293–319. [CrossRef]

19. Boy, D. Le grenelle de l'environnement: Une novation politique? *Rev. Française Adm. Publique* **2010**, *134*, 313–324. [CrossRef]
20. Legifrance. *Légifrance Décret N° 2012-557 Du 24 Avril 2012 Relatif Aux Obligations de Transparence Des Entreprises En Matière Sociale et Environnementale*; Legifrance: Paris, France, 2012.
21. Tronquet, C.; Grimault, J.; Foucherot, C. Etudes Climat. January 2017, p. 32. Available online: https://www.i4ce.org/wp-core/wp-content/uploads/2017/01/Potentiel-et-d%C3%A9terminants-de-la-demande-volontaire-en-cr%C3%A9dits-carbone-en-France-9.pdf (accessed on 5 November 2018).
22. CNPF Forêt et Carbone. Available online: https://www.cnpf.fr/n/foret-et-carbone (accessed on 4 September 2020).
23. Gren, I.-M.; Aklilu, A.Z. Policy design for forest carbon sequestration: A review of the literature. *For. Policy Econ.* **2016**, *70*, 128–136. [CrossRef]
24. UNFCCC. *Kyoto Protocol–Reference Manual on Accounting of Emissions and Assigned Amount*; UNFCCC: Bonn, Germany, 2007.
25. IPCC Change. *Climate Change 2014: Mitigation of Climate Change*; Edenhofer, O.R., Pichs-Madruga, Y., Sokona, E., Farahani, S., Kadner, K., Seyboth, A., Adler, I., Baum, S., Brunner, P., IPCC, Eds.; Working Group III Contribution to the Fifth Assessment Report of the Intergovernmental Panel on Climate Change; Cambridge University Press: Cambridge, UK; New York, NY, USA, 2014; ISBN 978-1-107-05821-7.
26. Pretzsch, H.; Forrester, D.I.; Rötzer, T. Representation of species mixing in forest growth models. A review and perspective. *Ecol. Model.* **2015**, *313*, 276–292. [CrossRef]
27. Biber, P.; Borges, J.; Moshammer, R.; Barreiro, S.; Botequim, B.; Brodrechtová, Y.; Brukas, V.; Chirici, G.; Cordero-Debets, R.; Corrigan, E.; et al. How sensitive are ecosystem services in European forest landscapes to silvi-cultural treatment? *Forests* **2015**, *6*, 1666–1695. [CrossRef]
28. Brockerhoff, E.G.; Barbaro, L.; Castagneyrol, B.; Forrester, D.I.; Gardiner, B.; González-Olabarria, J.R.; Lyver, P.O.; Meurisse, N.; Oxbrough, A.; Taki, H.; et al. Forest biodiversity, ecosystem functioning and the provision of eco-system services. *Biodivers. Conserv.* **2017**, *26*, 3005–3035. [CrossRef]
29. Mathias, J.-D.; Bonté, B.; Cordonnier, T.; de Morogues, F. Using the viability theory to assess the flexibility of forest managers under ecological intensification. *Environ. Manag.* **2015**, *56*, 1170–1183. [CrossRef]
30. Barnaud, C.; Corbera, E.; Muradian, R.; Salliou, N.; Sirami, C.; Vialatte, A.; Choisis, J.-P.; Dendoncker, N.; Mathevet, R.; Moreau, C. Ecosystem services, social interdependencies, and collective action: A conceptual framework. *Ecol. Soc.* **2018**, *23*. [CrossRef]
31. IFN Inventaire Forestier National. Available online: http://inventaire-forestier.ign.fr/ (accessed on 27 January 2020).
32. Cashore, B.; Egan, E.; Auld, G.; Newsom, D. Revising theories of nonstate market-driven (NSMD) governance: Lessons from the finnish forest certification experience. *Glob. Environ. Politics* **2007**, *7*, 1–44. [CrossRef]
33. Sylv'ACCTES. Available online: http://www.sylvacctes.org/ (accessed on 10 July 2019).
34. Gusti, M.; Di Fulvio, F.; Biber, P.; Korosuo, A.; Forsell, N. The effect of alternative forest management models on the forest harvest and emissions as compared to the forest reference level. *Forests* **2020**, *11*, 794. [CrossRef]
35. Luyssaert, S.; Schulze, E.-D.; Börner, A.; Knohl, A.; Hessenmöller, D.; Law, B.E.; Ciais, P.; Grace, J. Old-Growth forests as global carbon sinks. *Nature* **2008**, *455*, 213–215. [CrossRef] [PubMed]
36. Stephenson, N.L.; Das, A.J.; Condit, R.; Russo, S.E.; Baker, P.J.; Beckman, N.G.; Coomes, D.A.; Lines, E.R.; Morris, W.K.; Rüger, N.; et al. Rate of tree carbon accumulation increases continuously with tree size. *Nature* **2014**, *507*, 90–93. [CrossRef]
37. Jonard, M.; Nicolas, M.; Coomes, D.A.; Caignet, I.; Saenger, A.; Ponette, Q. Forest soils in France are sequestering substantial amounts of carbon. *Sci. Total Environ.* **2017**, *574*, 616–628. [CrossRef]
38. Shin, S.; Yeo-Chang, Y. Perspectives of private forest owners toward investment in forest carbon offset projects: A case of Geumsan-Gun, South Korea. *Forests* **2019**, *10*, 21. [CrossRef]
39. Maraseni, T.N.; Dargusch, P. Expanding woodland regeneration on marginal southern queensland pastures using market-based instruments: A landowners' perspective. *Australas. J. Environ. Manag.* **2008**, *15*, 104–112. [CrossRef]
40. Fortmann, L.; Cordero, P.; Sohngen, B.; Roe, B. *Incentive Contracts for Environmental Services and Their Potential in REDD.*; The World Bank: Washington, DC, USA, 2014.
41. ONF. *Contrat d'objectifs et de Performance 2016-2020*; Office national des forêts: Paris, France, 2016.
42. Cinotti, B. Une gestion non rentable peut-elle être durable? *Rev. For. Française* **2003**, *55*, 7–24. [CrossRef]
43. Barthod, C. La Réforme des financements publics aux investissements forestiers. *Rev. For. Française* **2001**, *9*. [CrossRef]
44. Legifrance. *Légifrance Code Général Des Impôts–Article 238 Bis*; Legifrance: Paris, France, 2018.
45. Primmer, E.; Paloniemi, R.; Similä, J.; Tainio, A. Forest owner perceptions of institutions and voluntary contracting for biodiversity conservation: Not crowding out but staying out. *Ecol. Econ.* **2014**, *103*, 1–10. [CrossRef]
46. CNPF. CNPF–Centre National de La Propriété Forestière. Available online: https://www.cnpf.fr/ (accessed on 24 June 2019).
47. Baskent, E.Z.; Borges, J.G.; Kašpar, J.; Tahri, M. A design for addressing multiple ecosystem services in forest management planning. *Forests* **2020**, *11*, 1108. [CrossRef]
48. Gutierrez Garzon, A.R.; Bettinger, P.; Siry, J.; Abrams, J.; Cieszewski, C.; Boston, K.; Mei, B.; Zengin, H.; Yeşil, A. A comparative analysis of five forest certification programs. *Forests* **2020**, *11*, 863. [CrossRef]
49. Bernstein, S.; Cashore, B. Can non-state global governance be legitimate? An analytical framework. *Regul. Gov.* **2007**, *1*, 347–371. [CrossRef]
50. Perrin, M.; de Noblet-Ducoudré, N.; Strada, S.; Torre, A. Plaidoyer pour le développement d'une action clima-tique régionale: Aménagement de l'espace, usage des sols et conditions atmosphériques. *Norois Environ. Amenage. Soc.* **2017**, 29–45. [CrossRef]

51. Ostrom, E. Beyond markets and states: Polycentric governance of complex economic systems. *Am. Econ. Rev.* **2010**, *100*, 641–672. [CrossRef]
52. Benz, J.P.; Chen, S.; Dang, S.; Dieter, M.; Labelle, E.R.; Liu, G.; Hou, L.; Mosandl, R.M.; Pretzsch, H.; Pukall, K.; et al. Multifunctionality of forests: A white paper on challenges and opportunities in China and Germany. *Forests* **2020**, *11*, 266. [CrossRef]
53. Konadu, D.D.; Mourão, Z.S.; Allwood, J.M.; Richards, K.S.; Kopec, G.; McMahon, R.; Fenner, R. Land use implications of future energy system trajectories—The case of the UK 2050 carbon plan. *Energy Policy* **2015**, *86*, 328–337. [CrossRef]
54. Legifrance. *Légifrance Décret No 2018-1043 Du 28 Novembre 2018 Créant Un Label "Bas-Carbone"*; Legifrance: Paris, France, 2018.
55. Barthod, C. Qu'a apporté le débat sur la gestion durable à la foresterie française? Ce débat doit-il être dépassé? *Rev. For. Française* **2007**, *59*. [CrossRef]
56. Assuah, A.; Sinclair, A.J. Unraveling the relationship between collective action and social learning: Evidence from community forest management in Canada. *Forests* **2019**, *10*, 494. [CrossRef]
57. UICN France. *Les Solutions Fondées Sur La Nature Pour Lutter Contre Les Changements Climatiques et Réduire Les Risques Naturels En France*; IUCN: Paris, France, 2018.

forests

Article

Forest Protection Unifies, Silviculture Divides: A Sociological Analysis of Local Stakeholders' Voices after Coppicing in the Marganai Forest (Sardinia, Italy)

Giampiero Branca [1,2], Irene Piredda [2], Roberto Scotti [2], Laura Chessa [3,4], Ilenia Murgia [2,5], Antonio Ganga [2], Sergio Francesco Campus [2], Raffaella Lovreglio [2], Enrico Guastini [2], Massimiliano Schwarz [6] and Filippo Giadrossich [2,4,*]

1 Department of Humanities and Social Sciences, University of Sassari, Piazza Conte di Moriana, 07100 Sassari, Italy; gibranca@uniss.it
2 Nuoro Forestry School, Department of Agriculture, University of Sassari, 08100 Nuoro, Italy; Irenepiredda@gmail.com (I.P.); scotti@uniss.it (R.S.); imurgia@uniss.it (I.M.); anto.ganga@gmail.com (A.G.); sfcampus@uniss.it (S.F.C.); rlovreglio@uniss.it (R.L.); grossopossum@gmail.com (E.G.)
3 AREA Science Park, Padriciano 99, 34149 Trieste, Italy; laura.chessa@areasciencepark.it
4 Desertification Research Center, University of Sassari, Viale Italia 39, 07100 Sassari, Italy
5 Department of Architecture, Design and Planning, University of Sassari, Piazza Duomo 6, 07041 Alghero, Italy
6 School of Agricultural Forest and Food Sciences, Bern University of Applied Sciences, Länggasse 85, 3052 Zollikofen, Switzerland; massimiliano.schwarz@bfh.ch
* Correspondence: fgiadrossich@uniss.it

Received: 28 May 2020; Accepted: 22 June 2020; Published: 25 June 2020

Abstract: Today, a forest is also understood as a real social actor with multiple-scale influences, capable of significantly conditioning the social, economic, and cultural system of a whole territory. The aim of this paper is to reconstruct and interpret the population's perception of the silvicultural activities related to traditional use of forest resources of the southwestern Sardinian Marganai State Forest. The "Marganai case" has brought to the attention of the mass media the role of this forest and its silviculture. The research was carried out via semi-structured interviews with the main stakeholders in the area. The qualitative approach in the collection and analysis of the information gathered has allowed us to reconstruct the historical-cultural and social cohesion function that the forest plays in rural communities. The results highlight that the main risks concern the erosion of the cultural forest heritage due to the abandonment of the rural dimension (mainly by the new generations, but not only), with the consequent spread of deep distortions in the perception, interpretation, and necessity of forestry activities and policy.

Keywords: forest planning and management; rural community sustainability; ecosystem services; forest sociology; forestry in the media

1. Introduction

The social component plays a fundamental role in the forestry sector. Some authors say that the social component is determined by those "who in a certain place and at a certain time have the right to decide the land use" [1]. At the same time, the forest policy process implies several interests. Public, private, and non-profit associations can influence forest policy today [2].

The forest planner cannot limit himself to applying technical knowledge based on the stationary characteristics of the forest in which he/she operates but must carefully analyze the socio-economic aspects that link the forest to the community. Today, the participatory approach is a recognized tool in public forestry management in Europe and North America [3]. However, it is not always applied with the right weight by forestry technicians and is often limited to a few public meetings and only at the information level.

The analysis of the perception of forestry issued by the community is an abundantly developed topic in this field of research [4–6] but the collected information is not always integrated into the planning processes. Furthermore, the participatory process can be adversely affected by difficulty in communication. The scientific sector often uses complex messages, which are not always attractive, and this can cause problems in communicating with the community. This difficulty, already presenting itself at the lowest level of involvement, can influence the whole process between planners and communities. A further problem is given by the widespread loss of historical memory, particularly evident in forest management. A striking example is given by the coppice silvicultural system. Coppice has been used for several centuries in the Mediterranean woods [7]. Although it has remote origins, the greater diffusion of the coppice system in Italy coincided with the population increase and the industrialization of the 19th century. In this historical period, coppice was connected to different wood assortments. The main products were firewood, coal, and sleepers, used in the construction of the Italian railway network. In the past, coppice has been associated with many uncontrolled practices in forests, such as grazing and agriculture. This association produced a strong environmental impact, building up the negative perception of the coppice system handed down to today, calling it "robbery forestry" [8,9].

The widespread interest in socio-environmental issues requires a new approach. Forest management has become particularly complex, as it must propose solutions capable of supporting socio-ecological systems. At the same time, new economic potentials are emerging: the correct management of forests should respond to new market needs [10]. In this perspective, collaborative activities between researchers in the sector, resource managers, and policy makers could generate interesting shared management strategies where the coppice management would take on an important role [11].

Today in Italy, the coppice woods are still widespread [12] and most of the coppice stands have reached or even exceeded twice the rotation age. Nevertheless, the recent proposal to re-introduce small-scale coppice management in the Marganai area has given rise to a heated debate at the regional and national level [13]. The influence of the media, due to the extreme synthesis the new media require and to some journalistic exaggerations, has led to a debate with actually no scientific grounds. The authority's intervention has led to the limitation of silvicultural operations in Sardinia as well as in other regions. A conflict between forest, forestry, and the landscape emerged. From a regional case, the forestry policy has been influenced up to the national scale.

An increasing number of research projects highlight an approach even more influenced by social sciences with respect to forest planning and management. References [14,15] have shown that the socio-cultural elements supporting the decision-making processes increase the socio-economic–environmental sustainability of the management interventions and reduce social conflicts. However, sociological research in forestry usually apply a quantitative approach that, through highly standardized techniques, aims to detect the incidence of pre-determined indicators. This approach detects static data that does not allow the emergence of new information, extraneous to those foreseen by the indicators identified ex ante. This scenario generates reliable results but does not capture innovative elements. The adoption of a rigorous qualitative approach for the socio-economic surveys concerning forest contexts can contribute significantly to improving the political and operational choices regarding land management [16].

The aim of this work is to study the contradictions of the perceived conflict between forests and forestry, through the perceptions within the local rural community, of the role of the forest and its management. A qualitative approach in this context is essential: the aim is to let those contradictions emerge, not to verify a priori hypothesis concerning the heart of the problem. Not relying on a random sample and being limited to relatively few interviews implies taking the burden of the responsibility, which we acknowledge.

To provide readers with a more detailed framework of this study case, in the next chapter we will introduce a brief history of the Marganai State Forest from the 19th century to the time that generated the case.

1.1. Context: Study Area and Forest Management from the 1850s to the Present

Marganai State Forest is located in the south-western part of Sardinia (Italy) and extends in a single body of about 3650 hectares on the Linas mountain massif (Figure 1). It is mainly located in the municipality of Domusnovas (with more than 6000 inhabitants). Smaller portions fall in other municipalities: Iglesias (more than 26,000 inhabitants) and Fluminimaggiore (about 3000 inhabitants). These municipalities occupy an area of 2324 km^2. The northern side of the Marganai forest, about 450 ha, is included in the Natura 2000 network as a Site of Community Interest (SCI), "Monte Linas-Marganai ITB 041111". In 1979, the Marganai forest was bought by the Italian government. Today it is managed by the Agenzia Fo.Re.S.T.A.S. (Forest Agency of the Sardinia Region).

1.1.1. The 19th and 20th Centuries

Historically, the Marganai State Forest overlaps with the forested area that in the 19th century was known as Oridda County (Figure 1). The story of this area is representative for the wide mining area of Sardinia called "Sulcis-Iglesiente". Oridda County became government property after the law of 1835 and was granted almost entirely to the municipality of Domusnovas. After the abolition of the fiefdoms, some of the woods were sold through private negotiations. In 1857, Oridda was sold to Count Beltrami. The count, a businessman and seller of timber, bought Oridda County in 1857 and was the first to use the forest for production purposes at an industrial level. In 1864 he sold the property to a mining company interested in the exploitation of the subsoil. The ownership passed to several mining companies, including the Scarzella family business, which held it for 55 years, from 1896 to 1951. Furthermore, in 1856, the firm John Taylor & Sons, seated in London, floated a company to pursue mining interest in these "celebrated" lead mines. The mining activity also resulted in the intensive exploitation of the forest's resources [17]. The intense production of coal has continued for about a century and continued until 1970. The structure of the forest and the dense network of mule tracks and charcoal burner's posts (Figure 2) bear witness to this silvicultural history.

In the last decades of the 20th century, the Marganai State Forest has been managed with a conservation approach, due to the abandonment of rural areas and the decline in demand for charcoal. Some parts of the forest have been thinned, aiming towards the conversion to high forest, while others have been left to natural evolution. Coal production is only a memory of the past and the production of firewood in the area is now limited to very small private properties.

1.1.2. The 21st Century

The Marganai forest has been left mainly to natural evolution (no silvicultural operations) from the 1980s to today, resulting in holm oak coppice of 40 to 50 years old. In 2010, with the double aim of recovering the 150-year-old historical silvicultural practices and to favor a local, sustainable forest chain for firewood, the Forest Agency implemented a management plan for a small portion of the Marganai forest's surface, less than 600 ha. Of this area, 354 ha were allocated to the reactivation of the traditional coppicing system [18].

The present case study has developed from a series of key events. Two other important planning processes initiated short after the 2010 forest management had started. The Marganai forest was included in a larger operation: the drafting of a forest management plan for the 13 state-owned forests in Sardinia [19,20]. Beside this, the renewal of the SCI management plan "Monte Linas-Marganai ITB 041111".

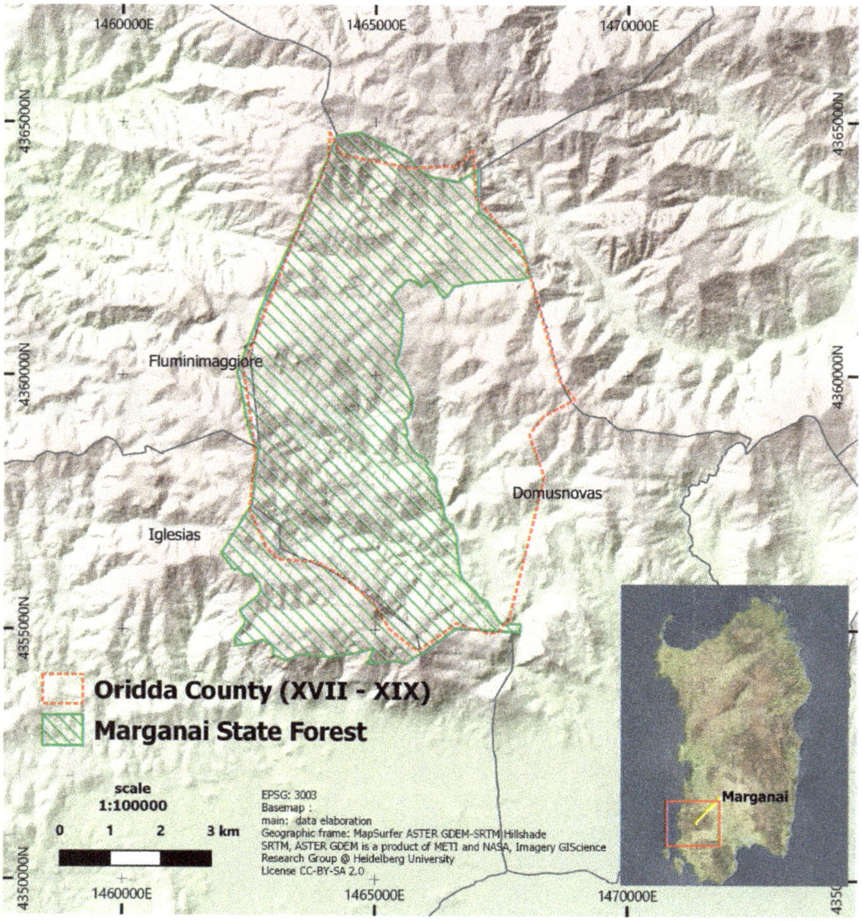

Figure 1. Geographical framework of the Marganai State Forest. Basemap: MapSurfer ASTER GDEM-SRTM Hillshade, a METI and NASA product, Imagery GIScience (License CC-BY-SA 2.0). The Oridda County border was processed from Real Corpo, 1842, Serie Mappe, Archives of State, Cagliari, Italy. Digitization by Techso SPA, 1999.

The larger planning operation activated interesting participatory processes for each forest. The first meeting for the Marganai forest was held in 2011, carrying out the information phase and presenting the planning project already underway.

The second meetings took place in 2013 and saw a moment of exchange between the local population, forest users, and planners. The interesting aspect is that the importance of the recreational and touristic function and that of the productive enhancement of the forest emerged in this context, as opportunities to revive the local economy and to meet the needs of the local market. The communities explicitly raised a request for firewood. The planners responded, identifying productive areas suited for the reactivation of the coppice system (personal communication).

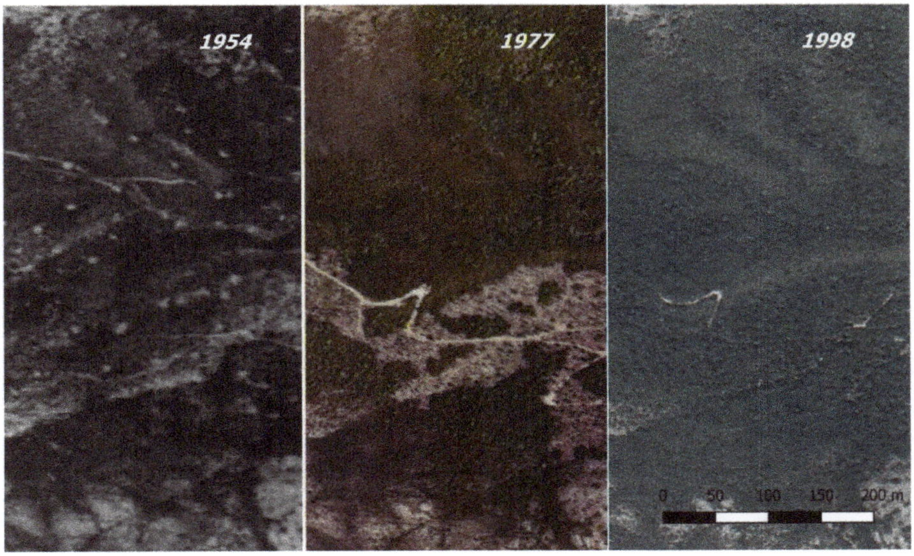

Figure 2. Sequence of images representative of the change in forest management, starting from the intense use in the post-war period to the present day. Mining activities and coal production shaped the landscape of this region and coppice wood dominated the management of this territory since the 19th century, in order to provide coal and lumber for industry, mines, and population. From the photo taken in 1954, it is possible to view the coal areas scattered throughout the territory. In 1977, coppice-with-standards areas at different stages of evolution are evident. Photos with permission from Sardinia Region.

In the same period, the professionals involved in the renewal of the plan for the Margani SCI expressed very strong concerns about the activation of coppice practices in the Marganai area [21]. During the participatory meeting, held in the Iglesias area in 2014, the professionals denounced the high risk of erosive phenomena triggered by such silvicultural operations [22]. However, Giadrossich et al. questioned the hypothesis of such risks in the Marganai forest [23].

The case has been amplified in the media and moved from local to national newspapers [13,23]. The media, following the criticisms highlighted by the professionals, further amplified the case, publishing titles and phrases like "The prehistoric forest that becomes firewood", "disastrous environmental consequences", and "deforestation of Marganai forest", just to mention a few.

The impact on the media has influenced local and national forest policies. The landscape superintendent at the regional level, pressured by the information disseminated by the media, investigated the case, and finally blocked the silvicultural works, launching a legal quibble. The planned developed of the identified productive area has been submitted to the authorities in accordance with the regulations in force for the Marganai forest. The interpretation of the legislation did not help the case and the length of the bureaucracy did not allow it to be resolved. Besides being marginally interested by a SCI, the whole forest is actually under a double bound (according to the following Italian laws: Legge 29 June 1939, n. 1497, "Protezione delle bellezze naturali", published in the G.U. n. 241 in the 14/10/1939; and D. Lgs. 22 Juanary 2004, n. 42, "Codice dei Beni culturali e del Paesaggio", published in the G.U. n. 45 in the 24/02/2004, s.o. n. 28). The procedures required for authorization are not straightforward. As some environmentalists raised the case with the administrative authorities, local operations have been blocked, waiting for clarifications. Moreover, the silvicultural operations in all the other coppice systems in Sardinia (and also in other regions) have been severely limited. The difficulty in interpreting the legislation has caused a blockage of the sector, stopping proposals for new forest management plans. All the people responsible for the planning process and the implementation of the Marganai silvicultural works have been reported.

The stop has also had economic consequences on the local and larger population. For the implementation of the silvicultural operations, the Forestas Agency had issued a call for a tender [24]. The contract was awarded to a company in the neighboring town of Domusnovas. The silvicultural operations started in 2010 but were halted by the authorities in 2014. The forestry company was sued. People lost jobs and the company lost the possibility of recovering its economic investment finally heading to its failure and to the closure of the company.

The "Marganai case" is still remembered by the Italian scientific community as an example of distorted information spread by the media and of conflicts both between forest perception and forestry and between effective legal competences and legislation interpretation.

2. Materials and Methods

2.1. Methodological Approach

We used the semi-structured interview as a detection tool. The interview, intended as a scientific tool for information collection, i.e., a conversation with a specific purpose, is the main tool in social research, providing reliable and comparable qualitative data [25,26]. Specifically, we used a qualitative approach that is characterized by the use of non-standardized procedures (or with a low level of standardization) for the collection of information on a limited number of cases taken as "typical and significant" [27]. This approach leaves an important margin of autonomy to the witnesses interviewed to expose their vision; it facilitates the emergence of unpredictable elements in the eyes of researchers, who are unrelated to the local communities and dynamics. The testimony collected through an interview is a representation of facts and personal experiences, interpreted and presented according to the point of view of a witness. Therefore, it is an unarguable—it cannot be considered "right or wrong".

The interview outline is divided into sub-topics to make possible an overview of the more specific investigative elements to be looked at in-depth during the information-gathering phase. The outline of the interview is the boundary that defines the scope of the investigation and the topics that must be addressed. It is a compass that allows the interviewer to follow the flow of thoughts of the witness, catching the unexpected but useful information, enhancing the originality, and "conducting the conversation" in accordance with the purpose set for the research [28]. The questions and insights are proposed to the witness, following the path of his thoughts, and stimulating the emergence of information, memories, personal opinions, and feelings [29]. In this research, the interview outline has been built around four macro dimensions, for each of which a set of sub-topics has been defined (Table 1).

2.2. Research Sample

The research sample consisted of an audience of 23 stakeholders (Table 2) defined as "qualified witnesses" [30]. This term indicates people who hold information on the topics useful for research purposes. In a qualitative approach the researcher carefully selects the witnesses based on their personal characteristics and on the skills/knowledge they hold [31]. The qualified witnesses are selected with a non-probabilistic procedure (reasonably chosen, [32,33]). This is justified by the fact that the personal characteristics of the witnesses significantly determine the success of the survey in terms of the quality, variety, and reliability of the information collected, decisively influencing the pursuit of the cognitive objectives of the research [34,35].

Table 1. Outline of the interviews.

Macro Dimension of the Interview	Main Tracks
History and traditional uses	What does the forest of Marganai mean in your life and in that of your community?
	How has the way of living the Marganai Forest changed in the local community?
Perception of resources	What is the territory that Marganai Forest refers to? (please sketch on the "silent map" that is on the table, your memory of the boundaries of the forest)
	How has the way of living the Marganai forest changed from a political/administrative point of view? (administrators only)
	How does the forest relate to the economic, political and social dynamics of the territory?
	In your opinion, what are the needs of the territory of reference?
	Are the material resources of the Marganai Forest an economic and work source for its territory?
	How can the Marganai Forest be valued from an economic point of view?
	What is your knowledge and opinion about the production and consumption of firewood in the territory?
Knowledge of coppice management	What information do you have about what happened as a result of the use of the Marganai's forest resources?
	What were your sources of information on this matter?
	What is your opinion about the use of the Marganai's forest resources?
	In your opinion, what is the relationship between the depopulation process and the possible dynamics of economic development?
	What relations do you have with the companies that have operated within the Marganai forest?
Future	How do you see the Marganai Forest in 10 years?
	Do you think you can do forestry in Sardinia? (administrators only)
	What business prospects do you see in the near future?
	Would you like to add something we haven't asked you?

Table 2. Composition of the research sample.

Witnesses	*n*
Public administration at different levels and organization	7
Companies in agroforestry, firewood, and timber sales	8
Naturalistic associations, hiking associations, museums, others social and cultural actors, and journalists	8
Total number of interviews	23

2.3. Interview's Analysis

Following the most common qualitative analysis methodology, the 23 semi-structured interviews were transcribed and analyzed through the identification of "meaning strings" (interviews phrases) obtained through the interpretative capacity of the researcher. Through analysis of the interview, we detected situations, behaviors, attitudes, and opinions, which are the result of the witness's personal vision of the reality under investigation [36]. The methodology led to the creation of interviews with very different contents, expressing what each witness deemed appropriate to communicate. The breakdown of the interview tracks into sub-topics (Table 1) and sub-dimension made up the storytelling of the interview, even if with different and original shapes, which are determined by the age, the skills, and the propensity to speak in front of a stranger. We highlighted significant witnesses' phrases that support homogeneous information categories (H.I.) and used these categories as a path for the discussion. The final product of the analysis consists of a series of coded information useful for teasing out the vision of the world of each witness.

To protect the privacy of the witnesses, the interview extracts were reported anonymously through the assignment of a progressive alphanumeric code randomly distributed (witnesses from W-01 to W-23). Each respondent has signed a release to the processing of personal data and the use of the testimony collected [37]. Interviews were carried out from June to September 2018.

3. Results and Discussion

The analysis of the interviews has produced an archive of information, rich in suggestions and useful for explaining the meaning of the communities' social action and for reconstructing and defining the role of the Marganai forest as a social actor capable of coexisting, in a symbiotic way, with the socio-economic system [38]. The relationship among the environmental, cultural, and economic components creates socio-cultural dynamics whose repercussions are considered far more important and significant than the value attributed to each single component of the relationship [39,40]. In fact, collected testimonies constantly show the awareness of living within a "community dimension" that embraces environmental, cultural, socio-economic, and historical elements. Results highlighted the presence of some central and recurring themes, summarized in the perspectives of the analysis and argumentation:

- Historical and identity function of the forest.
- Intergenerational cultural erosion.
- Socio-economic dimension of forestry.
- Perception of silvicultural activities.

In the following sections we discuss each of these four recurring themes, crossing all testimonies.

3.1. Historical and Identity Function of the Forest

The first element that unites the testimonies concerns the identity function of the Marganai forest for the local populations. Table 3 shows the homogeneous information collected during the investigation, which has clearly revealed the existence of this link and the important role that the forest plays in the community path of collective identity construction (Table 3). Over the centuries, the Marganai forest has played a central role in the history of local communities: from Roman domination until just a few decades ago the documented mining activities that led to the exploitation of other environmental resources, thereby influencing the local economy [9,17]. The presence of the forest of Marganai, both as economic and ecosystem agent, has inevitably conditioned the lifestyles and wellbeing of the population, determining the main features of the community identity. Within this social-cultural-environmental triangle, nature (the "non-human" element) has met the cultural dimension (social product through which man organized the material world) and the economy (social institution at the base of the social order) [41]. The interweaving of these factors is the result of a complex set of social relations, historical events, and socio-economic dynamics that have involved both the Marganai forest and the local populations. The link between the rural populations and territory is characterized by a sense of belonging that has a cultural base made up of a system of values and symbols linked to the land, the soil, and the territory [42,43].

Table 3. Examples of homogeneous information (H.I.) about the dimensional analysis of the historical and identity functions of the forest.

H.I.	I.C.	Witness's Statement
Collective identity	W-03	"It is an identity factor. Identity [...] concerns the sense of recognition of places, that is, the recognition of home, and the sense of landscape, both physical and human."
	W-13	"The Marganai is part of my personal history. As a child, I used the Marganai forest. [...] Marganai forest plays an educational role for us."
	W-14	"We felt the Marganai as something that belonged to the community."
	W-17	"The presence, even visual, of a massif of this type [Marganai's mountains] has influenced the modus vivendi of all the inhabitants of the territory."
Forest as a social actor	W-02	"From an historical-cultural point of view, it is an area that represents a point of recognition of the identity of the territory: local people have a point of reference to read their own story because the Marganai has always linked the relationships between local populations and the resources it has offered."
	W-11	"What does it represent for the inhabitants of Iglesias? Surely a green lung used over the years to make trips, to go hiking and to disconnect from the urban context."
	W-13	"I think every inhabitant of Iglesias has a piece of his history linked to the Marganai forest."
Resources of the forest system	W-09	"Long ago, people did not create too much trouble with their use of wood. Maybe they did so with intelligence because these are forests that have endured over the centuries. Evidently, there was a certain ability to work the forest, especially compared to today. [...] Perhaps the old men were ahead of many young people."
	W-14	"The territory lived in symbiosis with the needs of the urban world."
	W-19	"The woods are anthropized places that live in a symbiotic way with the man who exploits them and that makes them grow in a decent way and does not destroy them."

Interviews also describe the importance of the Marganai forest intended as a "social actor" (Table 3), socially and historically significant in the processes of the local communities' socialization to rural life (see also [44,45]). It is possible to re-define the symbolic, cultural, and social function of the Marganai forest thanks to the interpretative transition of the same forest from a "landscape element" to "social actor". This change in social function (from ecosystem to social actor) tends to define the forest as an active subject in the community that can condition social dynamics [46]. According to the interviewees' point of view, the environmental and ecosystem context exists and takes shape through the realization of a process of cultural interpretation that influences the perception of the forest for people who experience and give it symbolic meaning (see also [4,47]). The witnesses agree on the central role of the Marganai forest in relation to the processes of defining the community identity: the local population and urban population benefit differently from the resources of the forest system

(Table 3). Rural communities perceive the natural environment as an element characterizing the cultural identity, while the urban population perceives more a tourist–recreational (aesthetic) value, based on a conservative attitude. The "culture of silviculture" is linked to a "tacit knowledge" transmitted from generation to generation through implicit and traditional education paths. This culture is not the result of conscious learning or ideological imposition, but is acquired through practice [48]. The community path of transmission of these operational skills originates in a "practical sense" that is realized not through educational pathways but thanks to a practical, daily operational experience in which words assume a residual relevance: the "know-how" prevails over the "knowing that" [49,50]. It is a symbiotic path of accompaniment to the direct and personal discovery of the "know-how" that is realized within the rural communities and it is in sync with the natural environment. Times and places become variables that determine the success of the transmission/transposition of knowledge. To guarantee the transfer of technical and cultural skills, a constant relationship (over a long timeframe) between the rural and ecosystem dimension is necessary [51].

3.2. *Intergenerational Cultural Erosion*

The presence of some central elements in contemporary society is beginning to emerge clearly through the interview's analysis: history, culture, environment, and economy form a scenario in which divergent visions and interpretations are defined [40,52].

This is the case of the progressive loss of skills and knowledge related to the cultivation and management of the forest. Local communities have always lived in close relationship with the forest, but the changing needs (Table 4) of the communities have modified the dynamics of exploitation and cultivation of the natural environment. Today, we might say that different ecosystem services emerge. It all began with the spread of the urbanization process that consisted of the abandonment of the countryside, of rural centers, and of the outskirts in favor of the cities [53,54]. This phenomenon occurred during the first years of the twentieth century but showed a significant acceleration after World War II. Whereas in 1950, the inhabitants of the cities numbered only 29.6% of the world population; by the year 2000 the percentage had risen to 47%. In 2007, for the first time in human history, the urban population surpassed that of the countryside. Furthermore, the statistical projections indicate that in 2050 the urbanized urban population will be equal to 69% [55]. The progressive depopulation of rural areas is a phenomenon that in Sardinia significantly affects the entire regional territory, to the advantage of the main urban centers of the island. Specifically, the three municipalities in which this research was carried out show an average early decrease in population of 5.1% in the period 2001–2017. A significant trend that also has negative repercussions on the economic and labor market dimensions. Consequently, relations with abandoned rural areas are no longer defined and no longer perceived as elements capable of conditioning social and economic life [56,57]. The forest, therefore, becomes a place away from the daily socio-economic dynamics. Over time, the forest assumes the characteristics of a natural space useful for recreational activities. There would seem to be an intergenerational rift between old generations, who have experienced the forest as a part of their life experience, and new generations, urbanized, and therefore not socialized to the dynamics and knowledge typical of the rural context. The consequence is the progressive disappearance of the intangible memory of practical and traditional knowledge. This is a cultural capital that has always allowed rural populations to interpret natural dynamics and interact harmoniously with the cycles of the forest [58,59]. The reason lies in the fact that the recipients of this process of socialization to rural life no longer inhabit those places. The new generations are for the most part urbanized. Even those who still live in rural areas are culturally urbanized; that is, they are included in socio-cultural subsystems distant from the dimension of values and knowledge on which the symbiotic coexistence is based between the local communities and Marganai forest [60].

Table 4. Examples of homogeneous information (H.I.) about the intergenerational cultural erosion.

H.I.	I.C.	Witness's Statement
Abandoned rural areas	W-01	"People are no longer used to non-urban outdoor life. Everyone goes to the sea but the Marganai forest is little known."
	W-02	"The voices being heard are the ones that are not from those places ... it's the public opinion of city-dwellers, people who live far away from those places. [...] What emerges, however, is a public opinion disconnected from the specific context."
	W-05	"[The local population] was split in two: those who know the territory, and saw similar works, were happy; others, who have never been in Marganai but who have seen it written on Facebook, have sided against it. The town was split in two."
	W-15	"Unlike the inhabitants of Domusnovas and Fluminimaggiore, the inhabitants of Iglesias are more detached from the mountain. [...] There are people who know other areas of Sardinia better, [areas that are] more publicized, but who do not know what we have near here."
Environmentalist	W-02	"What is happening is a lack of rural culture, and forestry in particular [...] that made it possible to understand certain dynamics of the forest, agriculture, breeding ... To understand these dynamics in depth and then to understand and support the balance between needs of local populations and system sustainability."
	W-15	"We are losing this culture, it is fading. We should combine it with the natural heritage."
	W-19	"I fear that the municipal administrations are sons of this abandonment of the forest. It is a historical question: if I were the Mayor, and I knew nothing about the forest, I would not even know how to exploit it in a positive way."
Urban-centric perspective	W-03	"The rural areas are today seen as "a some-place" that is at the disposal of the city"
	W-13	"I have an opinion that sometimes clashes with the opinion of "Taliban naturalists". I live the forest, I agree not to rape it, but the forest must be usable for everyone. You must give everyone the opportunity to use it."
	W-19	"Everything is the result of ignorance. To relate to the forest as a tropical forest where there are natives who have never seen a human being ... This is the product of a very radical-chic attitude. The new generation of environmentalists knows nothing about the forest and approaches the forest of Marganai thinking that it is the Amazon."
	W-22	"A certain type of environmentalism, which I call "environmental Talibanism", has conditioned political choices."

Table 4. *Cont.*

H.I.	I.C.	Witness's Statement
Loss of cultural capital	W-06	"Because people are used to seeing deforestation in the Amazon forest. Because people now know everything about an African elephant or a Bengali tiger, but they do not know a partridge, they do not know what a hare is. This is the drama of Sardinians."
	W-09	"The environmental aspect... the fact that a forest is destroyed has been held up as a scandal. [. . .] There was a contrast between foresters, the municipal administration of Domusnovas and environmentalists, who do not look favorably upon use of the forest."
	W-10	"The environmentalist or the animalist is always listened to. I say to be cautious because they are not always right. But we must try to find common ground."

The greatest expression of this inter-generational fracture (Table 4) is the loss of the so-called instrumental skills (linked to the cultivation and culture of the forest) and the consequent interpretative distortion of the role and functions of "nature" in the eyes of "citizens". According to witnesses, the main problem concerns the loss of rural and forest culture that led to the inevitable "abandonment of the forest", both from a cultural point of view and from a political/administrative perspective. The "rural and forest culture", typical of the old generations, played an essential role in interpreting and better understanding the balance between the needs of local populations and the sustainability of the system in environmental and economic terms. The various forms of forest resource use are also affected by this decision-making process. Sometimes, however, the socio-economic system is not able to enhance the uses and values of the forest, indicated by our witness as "forest dynamics", linked not only to the exploitation of naturally occurring resources but also to agriculture and breeding [61].

When the socialization of the new generations to these cultural elements has ceased, the interpretative perspective on the role and function of natural environments has changed in favor of an urban-centric perspective. The term "Taliban" is often collocated with that of "environmentalist" (Table 4). The meaning tends to highlight the radicalization of environmentalist thought, disconnected from the constraints and opportunities of everyday social life. With this expression, our witnesses wanted to underline the dangerousness of anachronistic attitudes towards a nature that is interpreted as "virgin" or falsely defined "millenarian". This is the result of the interruption of the process of regeneration of the rural culture that led to the modification of the vision of the natural environment as "someplace that is at the disposal of the city", a vision founded on the idea that nature regulates itself and man must never intervene or interfere. This perspective originates in a post-industrial context according to which the natural environment is interpreted and experienced as a resource to be exploited without any limit, in favor of a de-regulated economic–industrial development [62]. Despite this, the witnesses say that the new generations are not able to recognize the important role of the Marganai forest and the strong interconnection that this has with the daily life of local populations (not only ecological but also socio-economic).

This vision tends to ignore all the silviculture activities that man has historically put into practice and transforms a culturally rooted instrumental competence into a practice of environmental devastation. The abandonment of the rural world led to the "loss of cultural capital" (Table 4) and loss of tacit knowledge, useful for balanced cohabitation with the surrounding natural environment. Furthermore, it has produced a generation of environmentalists who live in sociocultural dynamics and contexts far from the physical place where their ideological vision of nature takes shape, with subsequent obvious distortions. The urbanization of large masses of population has certainly contributed to the construction of a romantic vision of natural environments and of the idea that tree cutting is always wrong. An excessive radicalization of this thought could, however, lead to a conservative perspective, to the detriment of a more balanced forest management capable of considering the multiple values and uses of the forest [14]. In this debate, forestry plays a central role as a concrete expression of the instrumental and traditional skills handed down in the rural communities.

3.3. Socio-Economic Dimension of Forestry

According to the words of our witnesses, silviculture and, in particular, the coppice system are "millennial activities" practiced since the "dawn of time" with the aim of "cultivating the forest" to put it "in condition to grow and to live better": the coppice activities guarantee sustainability and increase natural renewal processes, contributing to the protection of local biodiversity and enhancing the multifunctional dimension of the ecosystem services of socio-economic interest [63–65].

Coppicing is commonly considered a traditional form of forest management, widely practiced in Europe and particularly in Italy at least since the Etruscan-Roman period [7,66]. The practice of coppicing is based on the ability of some species of trees to regenerate promptly, starting from their stock after cutting; this is an extremely simple principle that, based on a deep knowledge of the regenerative capacity and "resilience of the forest" (Table 5), has allowed, over the centuries, to exploit the forest resources in a balanced fashion [1,67]. The coppice has guaranteed to the local communities the regeneration of forest raw materials, i.e., firewood and timber, necessary for the subsistence of the population, both for the daily needs and for the local socio-economic system [68]. However, forest management has long been oriented toward maximizing the returns of a limited set of outputs.

On the other hand, the last decades have seen the spread of a romantic and symbolic view of the forest, not only because there has been an increase in sensitivity to environmental protection. As mentioned, this is due to the intensification of the urbanization phenomenon that has contributed significantly to the dispersion of the rural cultural heritage founded also on the knowledge of the ecological dynamics of the forest environments. The generational fracture that has characterized recent history has inevitably affected the relationship between time and places, compromising the socializing process and limiting the spread of traditional cultural elements. The critical issues to which we refer have particularly affected the instrumental skills related to the practice of coppicing. The recent generational fracture has led to the reduction of cultural competencies related to the management of forest resources. Public opinion was divided between those who continue to inhabit the rural area and those who base their vision of the forest on ideological elements, ignoring the symbolic and practical values that characterize the deep relationship between man and nature. The distance between these two positions is even more evident if the focus is on the coppice. In this case, the possession or absence of the instrumental skills related to the forest dimension make clear the attitudes for and against the cuts (Table 5) among the local populations.

Silviculture, and coppicing in particular, is based on the elaboration of traditional instrumental skills (linked to the rural dimension) within scientific paradigms widely recognized as valid. However, public opinion is still deeply divided on this issue, with consequent ideological conflicts that negatively influence the already precarious equilibrium that binds man to the natural environment.

Table 5. Examples of homogeneous information (H.I.) about the socio-economic dimension of forestry.

H.I.	I.C.	Witness's Statement
Resilience of the forest	W-04	"Before, the forest tended to be better than now. Now the cuts have been abandoned. One goes there and can almost not pass through. You only find holm oaks. [...] The forest must not die but we must try to make it rejuvenate over time. It must always be alive."
	W-05	"Before, when the forest was inhabited, it created income. It was lived in and gave lots of wood to the population. This too was "inhabiting the territory". Being present also means preserving."
	W-06	"If you leave a forest alone, over the decades you will have nothing but a holm oak forest [...] you will not find anything underneath, not even a blade of grass. If you periodically exploit the forest you have an economic yield from the wood that you collect, and, what's more, you're rejuvenating the forest."
Attitudes for and against the cuts	W-14	"Why does silviculture frighten us? It has always been there since the dawn of time, whereas before logs were used to roll the rocks to build the pyramids, today they are used to make firewood and for other uses. [...] The moment we limit its use, we no longer have any connection with the habitat of human beings."
	W-19	"I have seen other woods in Italy [...] that an ignorant person like me could judge as a primitive forest, but in reality, is the result of a "cultivation". "Millennial cultivation" by the monks who transformed it into that wonder that is now. So, I've got a slightly different idea of what a forest must be like."
	W-11	"My idea is that the forest is like the home garden and must be taken care of. We do not have to be radicals but if we want to use the forest, we have to make it usable. The holm oak forests must be pruned and put in condition to grow more and live well. It must be cleaned and the undergrowth must be made accessible. Otherwise, it starts to deteriorate."

3.4. Perception of Silvicultural Activities

What is now denounced as "violence against the forest" (Table 6), in the past was interpreted as a symbiotic activity between man and nature. The set of activities concerning the care and cultivation of the forest are considered one of the most important forms of "culture" for the rural communities because of the close interdependence link between them and the forest environment. In fact, rural communities have always drawn food and energy resources from forestry activities. At the same time, the cultivation of the forest guarantees its vigor and wellbeing, making it flourish and thrive [69,70]. However, the intertwining of social phenomena, such as "generational fracture", "urbanization of the population", and "depletion of rural cultural heritage", has progressively influenced the dissemination of information on forest activities. With the lack of socialization paths in the community, the transfer of cultural skills is now achieved through the Internet and social networks [71]. Public opinion is constructed starting from that portion of the population most interested in particular themes, which interacts with the flows of information transfer and which determines the orientation of information contents in ideological terms [72]. The spread of hostile positions and prejudices about forest activities would, therefore, seem to be the result of a "communicative action" of this type [73]. The social communication processes have thus led the population to form a "public opinion" (Table 6) divided into two fronts: on the one hand, that based on technical information and favorable to a traditional use of forest resources; on the other, a public opinion influenced by social networks, which convey information associated with the devastation and indiscriminate exploitation of the natural environment.

From testimonies emerge the opposing positions to the practices of silviculture. They are the result of the synergistic intertwining of trends with different characteristics. Some of these are related to the lack of knowledge that witnesses have of forest life, a dimension perceived solely for recreational purposes. Other testimonies would seem to be deeply influenced and conditioned by the specific communication dynamics that orient the communication flows (Table 6) towards ideological tendencies contrary to silviculture. Public opinion is oriented towards an interpretation of forest management activities as harmful for the balance of ecosystem services and environmental regulation (climate, water, natural disasters, pollination, and infestations) [74].

The information about silvicultural activities carried out in the Marganai forest has been found by our witnesses mainly via online communication: social networks, newspapers, or blogs. The information conveyed through social media is hardly submitted to the control of reliability and scientific validity (Table 6) before dissemination. The result is possible exploitation and distortions for ideological purposes ([13], among others). In the specific case of the cuttings of the Marganai forest, this was possible because the public institutions would not seem to have adequately considered the importance of creating a participatory information path with respect to these issues.

Table 6. Homogeneous information (H.I.) about the perception of silvicultural activities.

H.I.	I.C.	Witness's Statement
Violence against the forest	W-01	"Even those who have studied silviculture for forty years tell us that the 'problem of cutting trees' in recent years is worsening because there is less and less historical memory of country life."
	W-02	"There are difficulties in the management of the forest because people who work in it, and are therefore aware of that cultural heritage that allowed us to understand the balance between natural resources and local populations, are few and isolated. They are struggling and increasingly relegated [ignored] and do not have time to give strength to their voice in the matter"
	W-14	"[Before] the territory lived in symbiosis with the needs of the urban world. [Now] the metropolitan citizen, who has all his comforts, [. . .] wants all forests to be virgin forests."
Public opinion	W-01	"Public opinion associates the cutting of the forest to the desert of Africa. So, cuts are equal to desertification and desert. [. . .] The cutting of the forest is a cyclic phase; it is a cultivation cut that is used to rejuvenate a forest that is not a natural forest."
	W-05	"[Public opinion] was divided into two. Those who know the area and have seen similar jobs were more than happy. Others, who have never seen anything on the spot but who have seen it written on Facebook, have lined up against. The country split into two."
	W-15	"What the public opinion knows is that this cutting has been done in an inconsiderate way . . . And surely it came out that they were destroying the forest."

Table 6. *Cont.*

H.I.	I.C.	Witness's Statement
Communication flows	W-09	"We read it especially on a site that launched this thing. [...] In fact, I do not know how the situation is. However, there was a conflict between foresters and the municipal administration of Domusnovas, and environmentalists who, as said the commissioner, do not look favorably at the use of wood."
	W-13	"There was a debate on Facebook and in the newspapers. They said: "They cut half of the Marganai forest". But it is only a part, it is certainly not half a forest. Unfortunately, with social media things are swollen to excess."
	W-14	"Social networks have combined damage with news sharing. Many people who shared that news were people who do not know the forest. Maybe even locals who you see in the square 24 h on 24 but who have never set foot on the Marganai [...]. But they were ready to sentence on the cuts."
	W-19	"What I know is that I have read online newspapers. I informed myself and that made me lean towards this position: "There is someone who has economic interests to destroy the forest". This is the image given by the newspapers."
Reliability and scientific validity	W-13	"Information is what is missing from the institutions. What's on the internet is not information, it's something else. The information is the official one of those who have decided these things: Mayor, Regional Administrator, and all the interested people. They should have said, "This is what is going to happen in Marganai". And I think that people who have skills and who know the mountains could also have agreed with this information."
	W-15	"At times alarmism is caused by not knowing what is happening. Surely, there is a lack of information."
	W-16	"The cuts have not been explained to the population. What has triggered the uprising is that this cut has not been explained. [...] From my point of view, information has not passed on or, if it has been given, it has not been enough. [...] I am convinced that if they had given more complete information, many would have been less opposed. If they gave guarantees there would be fewer problems."
	W-19	"If the municipal administrations had thought of involving stakeholders, perhaps involving them in some collective assembly and explaining the development plans, maybe we would not have arrived at this point."

Witnesses highlight the lack of a role of guidance and information on the part of public institutions, a role that could have mediated and prevented ideological contrasts [75,76]. Planning a participatory path of stakeholder engagement can be useful on several levels. From a cultural point of view, it could promote the rediscovery and diffusion of cultural elements linked to forest cycles. It could also contribute to worsen the generational fracture created by the abandonment of the rural dimension. From a socio-economic point of view, coppice forest management fosters economic activity related to the processing of wood and, therefore, produces wellbeing also in terms of social protection [77].

If public institutions do not take responsibility for promoting actions for the creation of a "well-informed public opinion", this informative space is occupied by other non-institutional subjects that can influence and direct public opinion towards values that are distant both from the rural culture dimension and from the scientific sphere.

4. Conclusions and Final Discussion

The qualitative approach has allowed us to reconstruct the perceptions of the local population concerning the forest and the connected socio-economic-cultural dynamics. The analysis of the information highlighted the presence of some significant elements, constantly present in the reference bibliography: human-nature interactions; the value of the witness of the stakeholders; and the perception about forest uses [78].

Firstly, a central role of the Marganai forest emerges in terms of its contribution to the community identity, its history, and relational network: the local population still recognizes itself in the forest. Identity, history, and social relationships are elements of a cultural dimension that can contribute to safeguarding the relational dimension between man and nature from possible fractures induced by "hypermodernity". Hypermodernity is taken to mean to all those processes that could potentially transform the rural environment into "non-places", into physical aseptic spaces (from a socio-cultural point of view) where people do not create relationships, do not produce history, and do not recognize each other in a shared identity [79]. In this scenario, people enjoy goods or services (environmental) individually and sporadically over time. The urbanization process strongly undermines the resilience of rural communities and leads to "cultural fragmentation".

Equally important is the stakeholders' perception of traditional uses of the Marganai forest's resources: the loss of identity of the link with the forest dimension undermines the intergenerational transfer of knowledge and practical skills related to forest cultivation. Intergenerational fracture is the result of the progressive loss of a community identity through which people have always interpreted the places of everyday life and built up stable social relations over time [60,80]. This fracture has negative repercussions on the local economic dimension that is historically linked to the traditional uses of forest resources. The "common knowledge" has been eroded by postmodern transformative processes with a consequent shift to the sphere of purely scientific "technical knowledge" [81]. The capacity of people to evaluate competently and silently any forestry intervention is also lost if cultural capital is no longer the result of knowledge handed down and experienced, or the result of practical experience, within a multidimensional perception of the natural environment in which the community lives. Emotionality fosters a vision of the natural dimension in which human interventions that can modify the natural environment (even if historically always implemented) are interpreted in a negative way and therefore hindered, or condemned, by the less informed public opinion.

This research highlighted also that if the population is adequately informed about silvicultural activities, a greater awareness of the complexity of the problems in the management of the community material (natural) and intangible (cultural) goods emerges (see also, e.g., [82,83]). The institutions responsible for governing the interventions should play a central role in an information process aimed at promoting the participation of citizens and the assumption of shared responsibility through effective planning. For example, they could involve citizens in informative paths on silvicultural activities, also putting the accent on the cultural–traditional, as well as scientific dimension, considering all the available options within the sustainability framework, from most strictly conservative prospects, limited to monitoring, up to the most productive ones like coppicing. It is widely believed in forestry that efficient and effective management of environmental resources should take into account and enhance the information and participatory processes that actively involve local communities. Participation creates awareness and shared knowledge that reduces the distance between "urbanized citizens", bearers of a romantic vision of the forest environment, and rural communities, holders of values and traditions that derive from the daily interaction with the physical resources of the forest. The balance between social, economic, and environmental needs produces tangible benefits and responses to the needs of the territory that can be addressed through participatory process [83].

The results of the research suggest the need to study the knowledge of silvicultural activities via an approach devoid of ideological connotations. The emotional dimension, linked to the perception of environmental risks [72], if combined with pragmatic elements of rural cultural capital, can lead public opinion to the creation of original insights about future socio-economic development. A multidisciplinary approach to the analysis of the socio-economic dimensions can help to develop new prospects for the forest management planning and decision-making process [84]. Finally, a holistic silvicultural model, which also includes the participative process, can safeguard different ecosystem services of the Marganai forest.

One last comment should be reserved for the methodological aspect. During this work, where sociologists and foresters collaborated closely, getting to understand each other's world, it became evident that social research in a forestry context cannot be conducted without adequate basic training, especially for semi-structured interviews. The person doing this must have a specific competence and mastery of the dynamics of conversation, which are often not part of the forester's cultural background. For example, with the exception of the question stimulus of departure (which is always the same and presented identically to all witnesses), the questions rarely follow the order indicated in the diagram [85]. The structure of the topics or questions of a semi-structured interview is not binding, neither in the formulation nor in the presentation order [86]. It is characterized by a high degree of discrepancy in the stimuli to which the witness is subjected. This allows unexpected but essential information to emerge helping to complete the reconstruction of the social phenomenon under investigation. The hope is that sociologists and foresters can collaborate more systematically in land and forest planning.

Author Contributions: Conceptualization, F.G., L.C., G.B., I.P., R.S., A.G., S.F.C., E.G., I.M., R.L., M.S.; methodology, G.B., L.C., I.P.; formal analysis, G.B.; investigation, G.B.; data curation, G.B.; writing—original draft preparation, G.B., F.G., I.P.; writing—review and editing, F.G., I.P., R.S., L.C., M.S., I.M.; visualization, A.G., I.M.; supervision, F.G.; project administration, F.G., L.C.; funding acquisition, F.G., L.C. All authors have read and agreed to the published version of the manuscript.

Funding: This research was funded by Regione Sardegna (Italy), co-funded by European Regional Development Fund (ROP Sardegna ERDF) [project code SULCIS-820965]. Project title: "Environmental and socio-economic sustainability of coppice forest utilization in Marganai".

Acknowledgments: The authors thank the project partners for their important contribution in the realization of the research. We are immensely grateful to all the respondents of the survey for their availability and for their precious testimonies. We would like to thank Andrea Vargiu and Mariantonietta Cocco of the Department of Humanities and Social Sciences, University of Sassari, for their helpful suggestions in the development of the qualitative approach of this sociological research. A special thanks to Elisa Marras for drawing the graphical abstract. We also thank Ludmila Ribeiro Roder, Francesca Angius, Matteo Piccolo and Gonario Sirca for their active support to the Marganai project. Finally, the municipality of Iglesias for their willingness and logistic support for the final conference of the project.

Conflicts of Interest: The authors declare no conflict of interest. The funders had no role in the design of the study; in the collection, analyses, or interpretation of data; in the writing of the manuscript, or in the decision to publish the results.

References

1. Piussi, P.; Alberti, G. *Selvicoltura Generale. Boschi, Società e Tecniche Colturali*; Compagnia delle Foreste S.r.l.: Arezzo, Italy, 2015; p. 432. ISBN 978-88-98850-11-2.
2. Bruna-Garcia, X.; Marey-Pèrez, M.F. Public participation: A need of forest planning. *iForest* **2014**, *7*, 216–226. [CrossRef]
3. Sectoral Activities Department International Labour Office. Public Participation in Forestry in Europe and North America. Report of the Team of Specialists on Participation in Forestry; Joint Fao/Ece/Ilo Committee on Forest Technology, Management and Training, Geneva, CH. 2000. Available online: http://www.unece.org/fileadmin/DAM/timber/joint-committee/participation/report-participation.pdf (accessed on 20 March 2020).

4. De Meo, I.; Paletto, A.; Cantiani, M.G. The attractiveness of Forests: Preferences and perceptions in a mountain community in Italy. *Ann. Res.* **2015**, *58*, 145–156. [CrossRef]
5. Schneider, J.; Ruda, A.; Kalasová, Z.; Paletto, A. The forest stakeholders' perception towards the NATURA 2000 network in the Czech Republic. *Forests* **2020**, *11*, 491. [CrossRef]
6. Paletto, A.; Giacovelli, G.; Grilli, G.; Balest, J.; De Meo, I. Stakeholders' preferences and the assessment of forest ecosystem services: A comparative analysis in Italy. *J. For. Sci.* **2014**, *60*, 472–483. [CrossRef]
7. Gabbrielli, A. Le vicende storiche e demografiche italiane come causa dei cambiamenti del paesaggio forestale. In *ANNALI*; Accademia Italiana di Scienze Forestali: Firenze, Italy, 2006; Volume 55, pp. 133–166.
8. Bernetti, G.; La Marca, O. Il bosco ceduo nella realtà italiana. In *Atti della*; Accademia dei Georgofili: Firenze, Italy, 2012; pp. 542–585. ISBN 978-88-596-1040-3.
9. Beccu, E. *Tra cronaca e storia Le vicende del patrimonio boschivo in Sardegna*; Carlo Delfino editore: Sassari, Italy, 2000; p. 432. ISBN 88-7138-196-3.
10. Marchetti, M.; Vizzarri, M.; Lasserre, B.; Sallustio, L.; Tavone, A. Natural capital and bioeconomy: Challenges and opportunities for forestry. *Ann. Silvic. Res.* **2015**, *38*, 62–73. [CrossRef]
11. Corona, P. Forestry research to support the transition towards a bio-based economy. *Ann. Silvic. Res.* **2014**, *38*, 37–38. [CrossRef]
12. Fabbio, G. Coppice forests, or the changeable aspect of things, a review. *Ann. Silvic. Res.* **2016**, *40*, 108–132. [CrossRef]
13. Stella, G.A. La Selva Preistorica del Sulcis che Diventa Legna da Ardere. Available online: https://www.corriere.it/scienze/15_settembre_07/selva-preistorica-sulcis-che-diventa-legna-ardere-1c366754-5524-11e5-b550-2d0dfde7eae0.shtml (accessed on 18 May 2020).
14. O'Brien, E.A. Human values and their importance to the development of forestry policy in britain: A literature review. *Forests* **2003**, *76*, 3–17. [CrossRef]
15. Farcy, C.; Devillez, F. New orientations of forest management planning from an historical perspective of the relations between man and nature. *For. Pol. Econ.* **2005**, *7*, 85–95. [CrossRef]
16. Schelhas, J.; Zabawa, R.; Molnar, J.J. New opportunities for social research on forest landowners in the South. *South. Rur. Sociol.* **2003**, *19*, 60–69.
17. Costa, R.L.; Canavera, E. *Domusnovas Dalle Origini al '900: Ricerca Storica, Documentaria, Bibliografica e sul Territorio*; Comune di Domusnovas: Domusnovas, Italy, 2001.
18. Airi, M.; Casula, A.; Asoni, G. *Piano Di Gestione Complesso Marganai-Ripristino Del Governo a Ceduo Su Aree Demaniali*; Personal Communication: Cagliari, Italy, 2010.
19. SardegnaForeste Notizie. Available online: https://www.sardegnaforeste.it/notizia/avvio-della-pianificazione-forestale-particolareggiata-nelle-foreste-demaniali (accessed on 18 May 2020).
20. Niccolini, M.; Perrino, M. Piano Forestale Particolareggiato Del Complesso Forestale "Marganai" Ugb "Marganai"-"Gutturu Pala" Relazione Tecnica. Available online: http://www.sardegnaambiente.it/documenti/3_68_20140701114331.pdf (accessed on 18 May 2020).
21. A.T.P "C.C.W.R "progettazioni e soluzioni ambientali, sviluppo equo ed ecosostenibile. Richiesta incontro tecnico: discussione su problematiche riscontrate su habitat forestali in area SIC MONTE LINAS-MARGANAI ITB041111 e limitrofe, 2015. Letter addressed to the "Sardegna Regional Forest Service", dated 11 Novembre 2014. Available online: https://gruppodinterventogiuridicoweb.files.wordpress.com/2015/01/richiesta-incontro-tecnico-ente-for.pdf (accessed on 28 September 2020).
22. Gruppo d'Intervento Giuridico odv. Foreste demaniali sarde e direttiva Habitat, un contributo del dott. Francesco Aru. Available online: https://gruppodinterventogiuridicoweb.com/2014/06/20/foreste-demaniali-sarde-e-direttiva-habitat-un-contributo-del-dott-francesco-aru/ (accessed on 18 May 2020).
23. Giadrossich, F.; Guastini, E. A critical analysis of Vacca, A., Aru, F., Ollesch, G. (2017). Short term impact of coppice management on soil in a Quercus ilex L. Stand in Sardinia. Land Degradation & Development 2019, 28, 553–565. *Land Degrad. Develop.* **2019**, *30*, 1765–1885. [CrossRef]
24. Piano Di Gestione Dei Tagli Boschivi Del Cf N °-15- Marganai– Ugb 1, Allegato A, Capitolato Tecnico Delle Condizioni Sotto le Quali è Posto in Vendita il Materiale Legnoso Proveniente dal Ceduo Semplice con Rilascio di Matricine del Bosco Sito in Località "Su Caraviu E Su Isteri" in Agro del Comune Di Domusnovas. Regione Autonoma della Sardegna, Ente Foreste della Sardegna. 2010. Available online: http://www.sardegnaambiente.it/documenti/3_233_20100204102403.pdf (accessed on 18 May 2020).

25. Berg, B.L. *Qualitative Research Methods for the Social Sciences*, 8th ed.; Pearson/Allyn & Bacon: Boston, MA, USA, 2007; p. 448.
26. Seidman, I. *Interviewing as Qualitative Research: A Guide for Researchers in Education and the Social Sciences*, 3rd ed.; Teachers College Press: New York, NY, USA, 2013; p. 162.
27. Jansen, H. The logic of qualitative survey research and its position in the field of social research methods. *Forum Qualit. Soci. Res.* **2010**, *11*. [CrossRef]
28. Valentine, G. Tell me about ... : Using interviews as a research methodology. In *Methods in Human Geography: A Guide for Students Doing a Research Project*, 2nd ed.; Flowerdew, R., Martin, D., Eds.; Prentice Hall: Essex, UK, 2005; pp. 110–126, ISBN-13: 9780582473218.
29. Vargiu, A. *Metodologia e Tecniche per la Ricerca Sociale*, 1st ed.; FrancoAngeli: Milano, Italy, 2007; p. 384.
30. Friedman, A.L.; Miles, S. *Stakeholders: Theory and Practice*; Oxford University Press: New York, NY, USA, 2006; p. 362.
31. Mitchell, R.K.; Agle, B.R.; Wood, D.J. Toward a theory of stakeholder identification and salience: Defining the principle of who and what really counts. *Acad. Manag. Rev.* **1997**, *22*, 853–886. [CrossRef]
32. Harrison, S.R.; Qureshi, M.E. Choice of stakeholder groups and members in multi-criteria decision models. *Nat. Resour. For.* **2000**, *24*, 11–19. [CrossRef]
33. Sarantakos, P. *Social Research*, 4th ed.; Palgrave Macmillan: New York, NY, USA, 2005; p. 536.
34. Ananda, J.; Herath, G. The use of analytic hierarchy process to incorporate stakeholder preferences into regional forest planning. *For. Pol. Econ.* **2003**, *5*, 13–26. [CrossRef]
35. Candrea, A.N.; Bouriaud, L.A. Stakeholders' analysis of potential sustainable tourism development strategies in Piatra Craiului National Park. *Ann. For. Res.* **2009**, *52*, 191–198. [CrossRef]
36. Schram, T.H. *Conceptualizing and Proposing Qualitative Research*, 2nd ed.; Pearson Merrill Prentice Hall: Upper Saddle River, NJ, USA, 2005; ISBN-13: 978-0131702868.
37. Regulation (EU) 2016/679 of the European Parliament and of the Council of 27 April 2016 on the Protection of Natural Persons with Regard to the Processing of Personal Data and on the Free Movement of such Data, and Repealing Directive 95/46/EC (General Data Protection Regulation). Available online: http://data.europa.eu/eli/reg/2016/679/oj (accessed on 18 March 2020).
38. Archer, M.S. *Making Our Way Through the World: Human Reflexivity and Social Mobility*; Cambridge University Press: Cambridge, UK, 2007; ISBN-13: 978-0521696937; ISBN -13.
39. Fukujama, F. Social Capital, Civil society and development. *Third World Quart* **2001**, *22*, 7–20. [CrossRef]
40. Wilson, G.A. Community resilience, globalization, and transitional pathways of decision making. *Geoforum* **2012**, *43*, 1218–1231. [CrossRef]
41. Weber, M. *Gesammelte Aufsätze zur Wissenschaftslehre*; TP Verone Publishing: Einbeck, Germany, 2013; p. 588. ISBN 9783863831936.
42. Goudy, W.J. Community attachment in a rural region. *Rur. Soc.* **1990**, *55*, 178–198. [CrossRef]
43. Hannerz, U. *Transnational Connections. Culture, People, Places*, 1st ed.; Routledge: London, UK; New York, NY, USA, 2001; p. 216. ISBN 9780415143097.
44. Geertz, C. *The Interpretation of Cultures*; Basic Books: New York, NY, USA, 1973; p. 480, ISBN-13: 978-0465097197.
45. Griswold, W. *Cultures and Society in a Changing World*, 2nd ed.; Pine Forge Press: Thousand Oaks, CA, USA, 2004; p. 193.
46. Corbett, M. Towards a rural sociological imagination: Ethnography and schooling in mobile modernity. *Ethnol. Educ.* **2015**, *10*, 263–277. [CrossRef]
47. Grieder, T.; Garkovich, L. Landscapes: The social construction of nature and the environment. *Rural Sociol.* **1994**, *59*, 1–24. [CrossRef]
48. Polanyi, M. *The Tacit Dimension*; The University of Chicago Press: Chicago, IL, USA, 2009; p. 128, ISBN-13: 978-0226672984.
49. Bourdieu, P. *La distinction. Critique sociale du jugement*; Minuit: Paris, France, 1979; p. 670, ISBN-13: 978-2707302755.
50. Bourdieu, P. *Raisons Pratiques. Sur la Théorie de L'action*; Seuil: Paris, France, 1994; p. 252, ISBN-13: 978-2020231053.
51. Phillips, M. Assets and affect in the study of social capital in rural communities. *Sociol. Rur.* **2016**, *56*, 220–247. [CrossRef]

52. Burke, P. *Sociology and History*; Allen & Unwin: London, UK, 1980; p. 116, ISBN-13: 9780043011157.
53. Wirth, L. *Urbanism as a Way of Life*; Chicago University Press: Chicago, IL, USA, 1964; pp. 60–83.
54. Brenner, N. *Implosion/Explosion: Towards a Study of Planetary Urbanization*; Jovis: Berlin, Germany, 2014; p. 573, ISBN-13: 978-3868593174.
55. United Nations. Department of Economic and Social Affairs Population Dynamics. Available online: https://population.un.org/wup/ (accessed on 27 January 2019).
56. Bell, M. The fruit of difference: The rural-urban continuum as system of identity. *Rural Sociol.* **1992**, *57*, 62–82. [CrossRef]
57. Murray, M.J. *The Urbanism of Exception. The Dynamics of Global City Building in the Twenty-First Century*; Cambridge University Press: Cambridge, UK, 2017; p. 434. [CrossRef]
58. Pongratz, H. Cultural tradition and social change in agriculture. *Sociol. Rur.* **1990**, *30*, 5–17. [CrossRef]
59. Mardsen, T.; Lowe, P.; Whatmore, S. *Rural Reconstructuring. Global Processes and Their Responses*; Fulton Publishers Ltd.: London, UK, 1990; p. 197. ISBN 1853461113.
60. Giddens, A. *The Consequences of Modernity*; Polity Press: Cambridge, UK, 1990; p. 200, ISBN-13: 978-0745609232.
61. De Meo, I.; Ferretti, F.; Paletto, A.; Cantiani, M.G. An approach to public involvement in forest landscape planning in Italy: A case study and its evaluation. *Ann. Silvic. Res.* **2017**, *41*, 54–66. [CrossRef]
62. Touraine, A. *Pourrons-Nous Vivre Ensemble?* Fayard: Paris, France, 1997; p. 395, ISBN-10: 221359872X.
63. Costanza, R.; D'Arge, R.; De Groot, R.; Farber, S.; Grasso, M.; Hannon, B.; Limburg, K.; Naeem, S.; O'Neill, R.V.; Paruel, J.; et al. The value of the world's ecosystem services and natural capital. *Nature* **1997**, *387*, 253–260. [CrossRef]
64. Farber, S.C.; Costanza, R.; Wilson, M.A. Economic and ecological concepts for valuing ecosystem services. *Ecol. Econ.* **2002**, *41*, 375–392. [CrossRef]
65. Felipe-Lucia, M.R.; Comín, F.A.; Escalera-Reyes, J. A framework for the social valuation of ecosystem services. *Ambio* **2015**, *44*, 308–318. [CrossRef]
66. Matthews, J.D. *Silvicultural Systems*; Clarendon Press: Oxford, UK, 1991; p. 296, ISBN-13: 978-0198546702.
67. Buckley, G.P. *Ecology and Management of Coppice Woodlands*; Chapman & Hall: London, UK, 1992; p. 336. [CrossRef]
68. Kneifl, M.; Kadavy, J.; Knott, R. Gross value yield potential of coppice, high forest and model conversion of high forest to coppice on best sites. *J For. Sci.* **2011**, *57*, 536–546. [CrossRef]
69. Kennedy, J.J. Conceiving forest management as providing for current and future social value. *For. Ecol. Man.* **1985**, *13*, 121–132. [CrossRef]
70. Schmithüsen, F. Multifunctional forestry practices as a land use strategy to meet increasing private and public demands in modern societies. *J. For. Sci.* **2007**, *53*, 290–298. [CrossRef]
71. Murphy, J.; Hill, C.A.; Dean, E. *Social Media, Sociality and Survey Research*; John Wiley and Sons: Hoboken, NJ, USA, 2013. [CrossRef]
72. Bourdieu, P. Public opinion does not exist. In *Communication and Class Struggle*; Mattelart, A., Siegelaub, S., Eds.; International General: New York, NY, USA, 1979; pp. 124–130. ISBN 0-88477-011-7.
73. Habermas, J. *Strukturwandel der Öffentlichkeit. Untersuchungen zu einer Kategorie der bürgerlichen Gesellschaft*; Suhrkamp Verlag AG: Frankfurt, Germany, 2001; p. 391, ISBN-13: 978-3518284919.
74. Nikodinoska, N.; Paletto, A.; Pastorella, F.; Granvik, M.; Franzese, P.P. Assessing, valuing and mapping ecosystem services at city level: The case of Uppsala (Sweden). *Ecol. Model.* **2018**, *368*, 411–424. [CrossRef]
75. Decaro, D.A.; Stokes, M.K. Public participation and institutional fit: A social-psychological perspective. *Ecol. Soc.* **2013**, *18*, 40. [CrossRef]
76. Folke, C.; Pritchard, L.; Berkes, F.; Colding, J.; Svedin, U. The problem of fit between ecosystems and institutions: Ten years later. *Ecol. Soc.* **2007**, *12*, 30. [CrossRef]
77. Tirivayia, N.; Nennena, L.; Tesfayea, W.; Mab, Q. The benefits of collective action: Exploring the role of forest producer organizations in social protection. *For. Pol. Econ.* **2018**, *90*, 106–114. [CrossRef]
78. Siegner, M.; Hagerman, S.; Kozak, R. Going deeper with documents: A systematic review of the application of extant texts in social research on forests. *For. Pol. Econ.* **2018**, *92*, 128–135. [CrossRef]
79. Augé, M. *Non-lieux. Introduction à Une Anthropologie de la Surmodernité*; Seuil: Paris, France, 1992; p. 150, ISBN-13: 978-2020125260.
80. Bauman, Z. *Liquid Modernity*; Polity Press: Cambridge, UK, 2000; p. 228. ISBN 0-7456-2409-X.

81. Zoppi, C.; Lai, S. Assessment of the regional landscape plan of Sardinia (Italy): A participatory-action-research case study type. *Land Use Pol.* **2010**, *27*, 690–705. [CrossRef]
82. Nijnik, A.; Nijnik, M.; Kopiy, S.; Zahvoyska, L.; Sarkki, S.; Kopiy, L.; Miller, D. Identifying and understanding attitudinal diversity on multi-functional changes in woodlands of the Ukrainian Carpathians. *Clim. Res.* **2017**, *73*, 45–56. [CrossRef]
83. Soe, K.T.; Yeo-Chang, Y. Perceptions of forest-dependent communities toward participation in forest conservation: A case study in Bago Yoma, South-Central Myanmar. *For. Pol. Econ.* **2019**, *100*, 129–141. [CrossRef]
84. Campus, S.; Scotti, R.; Piredda, I.; Murgia, I.; Ganga, A.; Giadrossich, F. The open data kit suite, a mobile data collection technology as an opportunity for forest mensuration practices. *Ann. Silvic. Res.* **2020**, *44*, 86–94. [CrossRef]
85. Bichi, R. *La Conduzione Delle Interviste Nella Ricerca Sociale*, 2nd ed.; Carocci: Roma, Italy, 2011; p. 216. ISBN 9788843038008.
86. Galletta, A. *Mastering the Semi-Structured Interview and Beyond: From Research Design to Analysis and Publication*; New York University Press: New York, NY, USA, 2013; p. 258. ISBN 9780814732939.

MDPI
St. Alban-Anlage 66
4052 Basel
Switzerland
Tel. +41 61 683 77 34
Fax +41 61 302 89 18
www.mdpi.com

Forests Editorial Office
E-mail: forests@mdpi.com
www.mdpi.com/journal/forests

www.ingramcontent.com/pod-product-compliance
Lightning Source LLC
Chambersburg PA
CBHW061554210926
51850CB00009B/2098
* 9 7 8 3 0 3 6 5 3 3 7 9 7 *